to Moses—
best adventures
J.R. Challacombe

Opening Hidden Frontiers

THE DRAGONS OF TIME

Also by J. R. Challacombe

CONQUEST OF GIANTS
Origins of the Conservation Ethic

The Dragons Of Time

Opening Hidden Frontiers

J. R. Challacombe

THE CLAIR STUDIES
BERKELEY
1994

OPENING HIDDEN FRONTIERS: The Dragons Of Time

THE CLAIR STUDIES, P. O. BOX 2336, BERKELEY, CA 94702

ISBN 1-886287-03-1
Library of Congress Catalog Card Number: 94-068854

DEDICATION

For C.C., my spirited
fellow time-traveller, who
shared the enthusiasm
of this inquiry.

THE CLAIR STUDIES
P. O. BOX 2336, BERKELEY, CA 94702

Dear Reader,

While the journey into wilderness is one's own, there is value in community. One step towards sharing and discussing the ideas herein might be communication through email. You can now reach me at the above address and by notes to "clairst@netcom.com."

No day is without its complications. The realization of this book was no exception. Two passages need amending. Page *xxi*: "Each sense of time brings with it its own powers, world-view, and dominant personality types." Page 86: "And with growing apprehension, the aging cityman finds himself being pushed ever closer to that 'emptiness' between yesterday and tomorrow."

Good luck on your adventures with the Dragons of Time.

J. R. Challacombe

Contents

A Note About Gender

Language is a system of imperfect symbols—representations corrupted and smudged by generations past. Writers must constantly worry with secondary and tertiary meanings and at the same time struggle to keep their presentations simple and effective. The terms, *man* and *mankind,* as used in these studies are generic. For every "he" there is an understood "she."

ACKNOWLEDGEMENTS

RECOGNITION AND APPRECIATION must be given to a number of individuals who became interested in this project and gave much encouragement and time.

Lou Schuster followed the study from the beginning over 20 years ago and kept pressing to get the material in print.

Erma Erman, Frank Smith, Diane Hourany, Ken Morton and Petra Kellers created an ongoing environment of shared concerns and values.

Ruth Shahinian devoted many hours to ferreting out errors in the manuscript of this essay and in the four yet to be published. Her efforts were heroic and invaluable. Whatever mistakes may yet be found are mine alone.

My mother supplied the impetus early on to question our culture and the individual's relationship with the not-known. Her mind was wide-ranging and investigative.

J. R. Challacombe
August 17, 1994

THE CLAIR STUDIES

PROLOGUE

SLOUCHING TOWARDS THE MILLENNIUM

THE MEANINGFUL USE OF TIME was defined for us two centuries ago in the so-called "Age of Enlightenment." Today in the western world, we continue our forefathers' quaint belief in equating the significant life with the "pursuit of happiness."

As a people we swarm in a continual, violent, heaping movement, forever casting aside the present and past in our quest for the illusive golden tomorrow. The engaged individual lives by wresting wealth from environments of limited defense. Success we measure pragmatically—by our ability to convert potentials into realities, resources into commodities.

This life of extracting values became, early on, the secular equivalent of what the religious in the 18th-Century proclaimed as the "unfolding of Divine Providence." America was heralded as the unoccupied, free land God had set aside for His chosen. Time was the playing field on which the Divine Purpose would be realized. When the Millennium occurred—that is, when King Jesus came riding his glorious rainbow-striped parachute back to earth—He would undoubtedly touch down in America.

Well, two hundred years have now passed.

Our forebears leveled the continental forest, hydraulicked mountains for gold, hunted otters to the edge of extinction. In the 1870s our grandfathers demonstrated to the Old World the benefits of ethnic cleansing. We ourselves poisoned thousands of acres with the chemical by-products of the Cold War. So what's next?

We stand today on the threshold of the 21st Century. Not only that, we shall soon enter the third millennium since the real Jesus decried the unquestioned material life. This twin event—the end of a failed, bloody century and the close of another tormented millennium—comes as a demarcation of sorts. Could it be a wake-up call?

A few will pause at this major juncture even as they would before an eclipse of the sun. In such rare moments our attention is forcefully directed to the grand cycles of Creation. Then it is that questions which rarely faze us may have a fleeting chance.

The pursuit of happiness? How shall we define it today? We no longer have a rich continent open and vulnerable to the callous exploitation we termed "progress." What then is our present measure of the meaningful life? And what about this business of winners and losers? Why should "winning" be our only option? Are there any worthwhile life gambits open to the "losers," or non-strivers in this game of commodity development and consumption?

Human beings are a curious, but dangerous part of creation. The phenomenon is not in the class of the lion or the whale or the ape. Mankind is more like fire, an earthquake, or an errant asteroid. The survival of the planet is in the hands of these wayward, violent, godlike creatures.

And now we are about to trip over a great juncture in man-made time and fall headlong into a "new" era. But will it be a continuation of the same? More ethnic cleansing? More rain forest destruction? More loss of species? More homeless and starving? More stunted, wasted lives? More losers?

What was that happiness we once had the right to pursue? Did the Heavenly Parachutist pass us by for a more likely planet? Can it be that we live in the twilight of the earth's existence?

Has the violence we call visual entertainment desensitized and petrified us in time? Are the hot ashes spewing nightly from a million electronic Vesuviuses burying us alive? What went wrong, America?

Do we want to know? Do we even care?

SURVIVING THE BURNOUT OF WESTERN CULTURE

EARLY MAN SAW CREATION as a great wonder. Each day he hoped it would unfold another layer of its mystery. He even spoke of creation, or god, as the "Great Mystery."

The psychology of origin man—the *urMensch* whom we prefer to dismiss as a primitive—allowed him to see the world and the universe as a marvelous, freeflowing complexity which would both instruct and sustain him. We know this for sure since a few excellent records and examples survived our genocide of these people in the late 19th Century.

Western psychology, on the other hand, is that of a caged animal. Not that of a wild animal impounded. But of a creature born and bred to the bars. To live outside a confinement would be unnatural for us and threatening.

We do not, of course, call this cage "a cage." We know it and speak of it lovingly as our culture. This is that complex of values, historical "facts," prejudices, working models, shared expectations and aversions which we inherit, add to, reinforce and pass on to our children. Narrowly defined, self-sustaining, self-perpetuating, this bloodless world-view makes up a tough shell through which we the living must interface and affect the real world.

Essentially, our culture—western culture—divides reality into good and bad, into light and dark, into we and they, into winners and losers, into insiders and outsiders, into resources and commodities. The minds born of this culture make the ultimate division of the world between the known and familiar and the undefined and threatening.

Strange to say, we in our western culture have no well-established psychology by which we can admit the unexplained and the challenging to affect and enrich our lives. The significance of this lack is monumental and catastrophic. To remedy this deficiency, Francis Bacon in the early 17th Century proposed to storm nature by a system of deductions—gathering phenomena and from them drawing logical conclusions. While this procedure sounds reasonable enough, it only allows us to perceive what we already understand. "Logical conclusions" are still within the domain of the established culture. Western thinking moves from the familiar to the familiar. From what is in hand as resources to what will in time be in hand when the resources have been transformed into images already defined. For us the "new" is not the mysterious discovered, but the familiar reinvented.

Instead of relating eagerly with the rich chaos which lies beyond the familiarity shell established by the culture, we are terrified of the unknown and unmapped. Any attempt by an individual or a few to introduce a radically new perception of the greater world or of man's nature—what is sometimes called a "paradigm shift"—faces opposition. The effort to expand understanding is met at first with a withering indifference. If that is not enough, violence follows. Every toxic known to man is used to kill off what threatens us from wilderness. Whether it be insects, plants, or strange ideas, we destroy that which is not already known and certified by the dominant culture.

The mind we use—that is, the set of values, attitudes, methods, goals we share daily—is not really our own. It is on loan to us from the culture. Although each of us is a living, flesh and blood creature, we rarely experience a genuine life. What lives through us is the culture and its goals and its value systems. At times we even kill and die for whatever it deems sacred. The culture is immortal; it knows not time. We, however, have but a few score years and then are gone. If we only continue the shell of the culture, then we do not truly live. We become carriers, or clones, of the culture. It lives, but we are the dead walking—a condition the Sphinx warned us of long ago.

Western culture has in our day reached a point of entropy. The barbaric massacres in Europe, in Africa, in Asia, in the Americas that we have witnessed this century evidence the real, ghastly nature of western culture.

For all the books on popular psychology it has not been proven that man is genetically violent and vicious. It can, however, be shown that the culture which underlies so-called "civilization" has always been dangerous. Particularly so now, as it plays out its death cycles.

Is there any escape from the burnout of the western world?

The mind we use daily is culturally-derived and tied directly into the dying. This mind that knows only alienation resides in our very blood and marrow, if you will. And it sucks the life experience from each of us. The mind of wilderness, on the other hand, the mind of the real-time, was ours at birth. It only waits for us to call it forth.

RESTART OF THE EVOLUTIONARY MIND

IT SHOULD BE CLEAR that the mind we use today and the mind that evolutionary man (*urMensch*) used are not the same. He perceived and inhabited a different world than that which we know and which we use to exhaustion.

Consider that early man lived and thrived on this planet for at least some three million plus years. Probably many more. The creature we immediately derive from was Cro-Magnon. We find him in deposits dated to some 30,000 years ago. So our original mind, the evolutionary mind—that is, the mind that related to the whole of creation, to the "chaos"—had a very long run. And the planet was not noticeably damaged by first man and the mind that nature provided for him.

Historical man appeared on the planet with his special values and science (good vs. evil; us against them; winners over losers; men versus women) at the end of the last Ice Age. This phenomenon of so-called male dominance has increased dramatically since about 2,500 B.C. As the new, western mind-set displaced the holistic, shamanistic world-views, so also appeared warfare on a large and barbarous scale.

In these past 10,000 years we have witnessed an accelerating degradation of the planet. But the most ugly, wholesale destruction of defenseless people and of the natural world was not in Greek and Roman times. Nor in the Dark Ages or Middle Ages. Difficult as those eras were they pale before our own times.

In the past century and especially in the past several decades, we have witnessed genocides upon genocides. We grow weary of them; our hearts become desensitized. At this very moment, millions are being removed or slaughtered in Bosnia and in Rwanda. Countless more are being systematically starved in the Sudan. To this elimination of entire peoples add the eradication of whole forest ecosystems, the displacement of ancient wisdoms, the poisoning and salinization of prime agricultural lands. All these were riches which took nature millions of years to develop. Once gone, they are gone forever.

The nature of this alteration—from the shamanistic, wisdom-seeking *ur*mind to the historical wilderness-destructive mind we know and use today—is without doubt the most fascinating and significant psycho-historical problem facing us. The seeds of that original mind are with us at birth. But western culture attacks them and substitutes its own psychology. It operates much like a small weak insect that inserts its eggs into a beetle. From the inside the larger bug is devoured by the larvae. When it dies, the small weak insects are ready to repeat the cycle.

How the shriveled, unnoticed remnants of our natural mind might be encouraged to restart in the midst of a hostile world is the purpose of The Clair Studies. As the name implies, the project's intent is to see the not-seen, to understand the unarticulated, to report the rarely discussed.

Thus far, four major essays have been prepared in the series. The first to be released is *Opening Hidden Frontiers.* Each essay stands alone. But when all are interlocked with the others, additional insights may be gained.

THE DRAGONS OF TIME

OUR STUDY IS BASED on a simple, intriguing idea. Time is an indeterminate—a mystery—which each man must define for himself. How the individual thinks of time and attempts to derive values from it will in turn define who he is and the outlines of his life. To avoid the challenge altogether condemns one to a living death.

It can now be shown that man in his existence on the planet has described and lived by *four* major definitions of time. Three of these created the grand epochs in man's history. Each time-set in its turn formed its own dominant culture, world view, traditions, societies and heroes. The mind which an individual used in one time-set differed radically from the minds of the others.

The effect of these several time-shifts on us today is that we have inherited *all* of them. But never before have we sorted them out. Consequently we are every moment at the mercy of their conflicting systems.

"Let us make an image of the soul," wrote Plato and went on to specify for his model a man, a lion, and a many-headed monster. "Now join them into one, and let the three somehow grow together.... Fashion the outside of them into a single image, as of a man, so that he who is not able to look within, and sees only the outer case, may believe the beast to be a single human creature."[1] Similarly, "The individual soul, like the State, has been divided by us into three principles." "There is one principle with which ... a man learns, another with which he is angry; the third has so many forms that we have not been able to give it a special name, but have denoted it by the general term appetitive, from the extraordinary strength and vehemence of the desires of eating and drinking and the other sensual appetites."[2]

Plato in *Laws* stated this theme of the internal division in man in its starkest simplicity: "All men are publically one another's enemies, and each man privately his own."[3] And a few lines later, "There is a war against ourselves going on within every one of us."[4] The campaigns of this 'intestine war,' he thought, would define the nature of man's compartmentalized mind.

In every century down to our own, we have seen artists, thinkers, and princely-types who have used this psychic triad to explain man's inner wound—the struggle of one part of his mind with another.[5]

Sigmund Freud, the well-known copyist, offered the trio of ego, id and superego as his own invention. Carl Jung spoke knowingly of opposites that had to be rejoined and of mysterious psychic archetypes. Berne formulated his psychology on a trio he described as Adult, Child and Parent.

So fixed has become this notion of man's split nature that well-funded psychophysiologists are even now probing man's 'older' brains with electric needles.[6] They hope to find the very 'seat' of these minds which Plato long ago speculated upon and Freud popularized. If the locus of the id/beast could be found—so goes the reasoning—the 'violence' some individuals express might be neutralized with a well-placed jab of a knife.

In this radioactive tangle of incompatibles—the restless beast, the reasoning self, and the 'angry old man'—we find some interesting historical parallels. The peasant who displaced the hunter so long ago thought of the hunter as a subhuman element. Enkidu in the *Epic of Gilgamesh* (c.2,000 B. C.) was a 'beast' of the field until he was introduced to the ways of the peasants and the early city. Only then did he become *human*. Intuition, clairvoyance, and the ecstatic visions which the hunter-shaman experienced seemed to the peasant the work of devils. In a like manner the cityman who displaced the peasant pictured the latter as less than fully developed. The man of the soil became the cranky old hayseed, bumpkin, oaf. His mind was focused on yesterday and a lot of meaningless rites, sacrifices, and mumbo-jumbo.

Modern cityman sees himself as none of the above. He is not the hunter who would leave the environment untouched for its sacredness. Nor is he the peasant who appears to think only of antique ways hallowed by millennia of repetitions. Cityman is the breaker of environments and the breaker of traditions. He is 'at war' with both the peasant and the hunter.

In this triad of historical conflicts it can be argued that we have the actual source of cityman's self-opposed mind. Each of us represents within his divided 'soul' the continuance of the original displacement of one mind by another mind and those two by still another. It is our thesis that what passes for our so-called tripartite achiever mind is in reality a layering of the several historical damages inflicted on man's origin mind.

The question that immediately arises—granted the above—is whether the divided can be rejoined and the parts married? This is the reasoning, fix-it approach once favored by Jung and others. But, we must ask ourselves, what is it we would patch or fix? What we have been using as mind, on closer examination, is more scar than viable tissue. Trying to make athletic muscle out of an organism's reaction to damage may never bring us close to the natural functions of the uninjured. If all this is so, then there can be no 'healing,' or 'marriage of opposites,' or other imaginable 'fixings.' The natural mind, which we have never really known or understood, can only be allowed to reassert itself. And that process will be closer to a second birth than anything else. But it will also entail, most certainly, the sharp diminution or even loss of the 'achiever control' we exercise over our vulnerable world.

We are artifacts of history. The mind we use day to day was never a simple *tabula rasa* but came already layered with whatever went before. We must deal not only with the personal story we add—our own stings and embarrassments—but also with the tragedies and promises of the millennia. The individual is mankind replicated.

In reality we swim in an unbounded time-wilderness. Early man delighted in this freedom. But that lack of borders and pathways frightens us today. We of the historical era are driven to structure the unmanifest and make it 'safe.' Therefore, we make use of time-sets which protect us—that is, alienate us—from reality and from each other. Since the time-sets we regularly turn to are historically derived, they are contradictory and even hostile to each other. As we shall see again and again, our everyday minds are ineluctably self-opposed and self-destructive.

Each of the time-mind-sets that cityman resorts to becomes another gigantic shadow flapping about in the dim, echoing cave of our unexplored self. Which of these dragons of time might be magical and friendly and which are threatening, which ones bring us bounties and which roar and belch fire, we are never sure. Here, we are on our own. Our culture and the 'psychology' it promotes are more hindrance than help.

Time and mind are not disparate, unrelated subjects. Rather, they are two sides of the same coin. Much of our so-called thinking—as we shall show at some length—is merely a processing of what we have accepted as everyday time symbols and values. Truly, one cannot speak intelligently of either time or mind without referring to the other. Each sense time of brings with it its own powers, world-view, and dominant personality type.

THAT HISTORY'S MANY CHANGING IDEAS of time have shaped the psyches of men and women is our subject. To explore beyond the margins of modern linear time is our purpose. Our mind—as you reflect on it, you will see—is little more than a time machine: our thoughts are continually passing from the past through the present to the future and back again. And good evidence abounds that many of our breakthrough discoveries about the nature of our world and our existence were made by individuals whose minds had for a moment a special experience of a time which is neither yesterday nor tomorrow. Is there any description, we desire to know, of Newton's state of mind when the idea of gravitation presented itself? What was happening when Professor Kekulé suddenly awoke with the vision of how the benzene molecule was structured? From whence does a great poet gain his insights into the hidden nature of relationships? And can we explain, other than as a miracle or freak of nature, the appearance of the man of wisdom, a Buddha, a Jesus, a Shankaracharya, a St. Francis, a Ramakrishna? Into these exciting matters—the relation between the superior mind and its direct experience of time—we shall delve. But we need also to discover how we ourselves are dependent on received ideas of time. And how these control our lives utterly.

From the earliest recesses of prehistory man has been concerned with the meaning of time. To notice the phases of the moon was to wonder about time. To count off the number of new moons until a baby was born was to be intimately involved with time. To look for the return of certain constellations as a signal to begin planting was to be governed by time. To give a name to each new moon of the year—as did the American Indians—was an attempt to order and structure time. To await the return of the salmon and the ripening of the cherries at the proper moons was to synchronize one's life with time and the greater life of the earth.

All men have in one way or another tried to define what time was, how it should be thought of, and what values might be realized as the sun rose and fell, as day followed day, and as one year dissolved into the next. Time has meant both what might be expected and what might be feared. And man has worked to assure the fulfillment of his intentions and to fend off time's unwelcome surprises. Wherever and whenever a man has put his forefinger on his pulse, his thoughts have inevitably been drawn to the meaning of time.

How the individual conceives of time shapes his life. His pathway—from birth to death—will be determined almost entirely by the central time formula into which he was born. Although it seems hidden, out of sight, this time parameter steadily and surely affects and determines the individual's personality and the quality of his life. The temporal canon by which we live might be compared to a pane of glass through which we observe the world. To us it seems transparent. We think we look at and handle reality. Seldom do we realize that the strains, colorations, and obscurities built into the glass distort the nature of what we deal with day to day. To see and think in real time—that is, without the window pane—is no simple matter.

Whatever the individual conceives time to mean will, of course, have been derived from his culture. Rare is the person who can think of time anew, apart from the matrix within which he has found himself. It is the prime function of a society to impress its meaning of time upon its individuals, generation to generation. This shared sense of the values and rewards to be derived from time gives a society its essential cohesion. It enables us to be recognized and to engage in conversations. Peoples who have different time values usually appear as aliens each to the other—members of another caste, or as nonhumans. In such a context an individual may now and then attempt to go his own way. But fight as he will, the time values built into his bones from childhood will warp him back into the common stream of his culture.

EACH OF US IS THE PRODUCT of *two* major historical forces. The first is an evolutionary push of several million years to create the body, brain, and mind of an alert, shifty, adaptable, hunter-explorer creature. The second is a peculiar drive of some 10,000 or more years to condition the natural mind and to separate this creature from his evolutionary roots and from his life in real time. The organism that we are has not changed that much, but the culture we inherit and pass on has gone through several radical permutations. The artificial and dangerous situation we now discover ourselves in is far removed from the evolutionary real-time continuum which produced our organism and its inherent sense of integrity.

In all that long span man has experienced, we have said, *four* time-sets. Among these major expressions have been numerous variations and overlappings. Essentially, however, there have been four: the time-mind of the early hunter-gatherer, the time-mind of the planter-herdsman, the time-mind of cityman (ourselves), and the unbounded time-mind of the one whom we shall call "the wilderness man."

The mind is the forum or stage on which man defines not only the space he occupies, but more importantly his sense of time and its meaning. The individual, however, does not ordinarily make these definitions for himself. They were made long before his birth and impressed on him by his particular culture.

Time for evolutionary man, the hunter-gatherer or "first Adam," was the everlasting present. He was well aware of yesterday and of tomorrow. But his emotional and psychic energies were focussed in intense and rewarding relationships with the immediate.

Time for the planter-herdsman, the peasant who with ruthlessness and violence displaced the hunter from his space, consisted of two elements of time, the present and the past. Again, he was fully aware of tomorrow. But his mind energies were bent towards yesterday. For in the past there lay, he was convinced, the great mysteries of life's origins. Creation had occurred at a specific, sacred moment in the past. And the peasant was continually trying to invoke that power event by his calendarian magic. In this way time might be mastered. Profane time (today) would by his rites and sacrifices be perfected through transformation into sacred time (that long-ago epiphany which had initiated the peasant world).[1] We still carry on many of these formulations in our observances of Easter, Christmas, the weekly Mass, and various other celebrations and anniversaries.

The Idyll of the Hunter lasted undisturbed down to about 10,000 B.C. It was then that the first radical shift in the meaning of time occurred with the coming of the cultivator-domesticator. And the subsequent epoch of the peasant as the 'newest' mind extended unbroken to circa 2,500 B.C. About here another radical shift in time-valuation happened. The even newer invention involved centering one's mind energies in tomorrow. Both the past and the present now became raw material out of which the future might be constructed.

The hunter had conceived that the present was holy and ongoing. The cultivator had believed that certain occasions in the past were sacred and must be repeated ad infinitum. The newest, or third mind-time set held, however, that the traditional calendar—whose sacred occasions and holy days had supposedly been set by divine powers—might be altered. Of course, only a god or one possessed of godlike powers might intervene to shape a nontraditional tomorrow. So radical was the new mind that it was strictly limited. The community endowed a specially created entity, one of their own, with the powers of their god—not only to protect the city against bandits, but also to handle the dangerous business of changing destinies and instituting new traditions.

This prerogative to tamper with set customs and to design a 'better tomorrow,' was held by one designated as priest-king.

This newest idea of time and its control by a resident divinity spread widely. And for some four millennia the presence of a wizard with powers over time-tomorrow was held essential to the well-being and orderly function of the properly organized community. And everywhere such kings claimed their powers to redefine traditions as theirs by "divine right." It was not until the mid 17th Century, in England, that a further radical change in man's conceptualization of time came along. At this point a limited democratization of the third mind—this special tomorrow-oriented mind of the king—commenced. With that, the doors leading into our modern era were blown off.

Praise has been heaped on this psychic model of the modern achiever. Our psychologies tout the goal-driven mind as being in charge of the facts, for making decisions, for turning 'waste' into value. For all their dazzle and powers of accumulation, however, such future-oriented minds are limited. They function only within the confines of a given culture—namely, within a milieu of time fictionalized as either past or as future. Such minds alienated from real time can at best be no more than second- or third-rate.

We have outlined man's three time epochs and the personalities which corresponded to each. But what of the fourth, the uncommon time-mind? Here we have what seems a sport of nature—the non-historical mind. For this unbounded mind has appeared in all three of the epochs which we have lightly sketched. Among the hunters the 'timeless' mind most often appeared as the shaman. In the long hunter tradition this mind was prominent and highly valued. But in formulating the time-mind parameters of the later resource-exploitive cultures, it never became a major player. During the era of the cultivators, the mind outside of time appeared as the prophet—the one who stood at the gates of the village and spoke the words of the Lord. In our more modern times such a mind has realized itself in the iconoclastic bard, the essayist-philosopher, and various others who have wandered in from 'the wilderness.' Loners and misunderstood though they may have been, their insights and discoveries changed the way we think of ourselves and of creation.

But how differently does this unhistoric mind conceive of time's meaning? In this fourth, or wilderness-discovery mode, the mind's energies are once again centered in the present. But the relationship is at another level than with the events of the chase. There is a merging in which the individual's mind is intensely questioning, challenging, listening for hints of the unspoken, unseen intelligence structuring the situation within which he finds himself. In addition to the available details, it would grasp the larger meaning that lies beyond the edge of what the mind already knows. Does such a mind sound like that of the breakthrough scientist? It is partially just that. But long before, it was the mind of the hunter-shaman.

Of this bardic, intuitive, discovery mind we would sing. For the genuine heroes of our kind are not those who wrap themselves in robes of power and lead armies or who collect artifacts and erect great institutions. They are instead the few minds that have pushed at the envelope of the familiar and have realized new perceptions of man's position and his potentials in this uncertain quest we know as life.

Each breakout realized by the intuitive hero becomes, to a degree, a breakout for all humanity. Liberation is not the freedom to aggrandize one's personality. It is, rather, that special experience, often called a "vision," whereby one's *understanding* suddenly expands beyond frontiers recognized by the dominant culture.

The most thrilling faculty of the human psyche is its capacity for voyaging free in real time. Then it is that the mind deals not with given "facts," but comes in direct contact with uncertainties and wonders. Only in real time can the intuitive, discovery experience break into the mind and flood it with an understanding not yet in books.

The fundamental mystery of how to reidentify with this exciting, real-time self cannot be resolved simply by adopting a 'positive attitude' or by making a series of determined choices. We presently assume an exact linear correspondence between action and effect: if I do this now, that will happen later. It is out of such naïveté that we believe we can choose and activate tomorrow. Life is, however, neither so simplistic nor so strained. It moves, but does not march. Our cityculture has denatured us to the rhythms and purposes of the actual world. And to find our way out of alien time and back into real time will require patience and close attention rather than action.

Everywhere we find rhythms and cycles. Like a pebble cast into a pond, each action sets a series of events, or disturbances, into motion. How they will interact with other ripples and waves we have no idea. Similarly, pull a boulder from a mountain. That effort dislodges another, and that another, and so on. We try to make life logical and controllable by positing reality in tomorrow. But life is today—cyclical, random, unpredictable, protean. We may be able to purchase a car in a linear fashion, so many payments over so many years. Yet life is ever a whirlwind of the unexpected and unplanned for. Who can foresee the exultation of a fresh discovery? Or indeed, who among us can predict when his or her son or daughter might perish? We must live our lives, whether we will or no, as ripplings of uncertainties. Each disturbance carries in it, like a seed, the potential for new insights. But these will occur only if the uncertainties are experienced directly in real time. For then, if at all, may some of life's secrets deign to reveal themselves before our astonished attention. At such moments the mind that was confused becomes "quick, ingenious,...subtle and sinewy."[7]

For those who like puzzles, this quest for the mind in real time is the ultimate mystery. Lewis Carroll gave us some hints of the difficulties in *Through the Looking Glass.* Melville allegorized the matter in the worldwide pursuit of a monster of the deep which destroyed the contrivances of overreaching men. And Robert Graves makes us live his compelling passion for the search in *The White Goddess.*

Clemenceau once observed, "War is much too serious a matter to be entrusted to the military." Likewise, the search for the real is far too important—even as is the act of breathing—to be left to others, especially to well-intentioned 'authorities.' We have many issues, injustices, and problems in our troubled world to address. But beneath all these alien time distortions lies the great question of what is man's *real-time* mind? And secondly, Where is it? Indeed, so central is this mystery, which each must puzzle out for himself, that one is driven to ask what else are we alive for—but to explore the ways of the mind in its age-old struggle with the meaning of time?

THE SIGN OF FOUR

THE NUMBER *FOUR*, interestingly enough, has played a considerable role in the mythology and magic of early peoples. All through the Bible there are references to the number four and its multiples. There were four rivers in Eden. Moses and the children of Israel wandered 40 years before reaching Jordan and the "promised land." Jesus spent 40 days in the Wilderness. The Christian cross is simply a latter-day symbol derived from any number of four-part creation symbols used throughout the Ancient Near East and India.

The aboriginals of North America also perceived that four had strange and marvelous properties. Black Elk spoke of the forces of creation as moving in four directions. The pottery, the shields and the shell gorgets which these people fashioned very often depicted the powers of the four directions.

Why does this figure keeps recurring? Who can tell? It's a marvel and we need not waste time here in trying to root out some cabalistic mystery. The "kicker," however, is that nature often does her business in four stages. And that wonder has been extensively studied and is called metamorphosis. The problem at hand for us is what are the real dynamics of man's thinking. Since he is one of nature's creations, should we be surprised if we find that *four* plays a role here also?

In this series of Clair Studies the movement of thought in several of these "minds" which man has used historically will be described as periodic. That is, the mental energy expended moves through specific, describable phases which taken together create a cycle. Each cycle is closed and self-fulfilling. To break free of any one of them is nearly impossible.

In the study now before us, *Opening Hidden Frontiers,* we shall examine closely the four periods of the Discovery Cycle. In a forthcoming essay, *Organize The Chaos*, the Achiever Cycle will also be analyzed. And in a third essay, *Upon The Fields of Time*, the breakdown of the hunter mind and the emergence of the peasant-vegetation cycle will be described. Finally, the fourth book in this group, *Life Of The Mind*, will trace the exciting and highly significant underground of wilderness thinking in western culture from the mid-17th Century to the present.

The thought pattern which now holds western cityman, and the planet, in thraldom is the Achiever Cycle. Because the Discovery Cycle described in the following essay is closely contrasted with this Achiever Cycle, a brief description here of the latter's four stages is in order. A full presentation of its historical derivation, its dynamics and its dangers will be forthcoming. The following are excerpts from *Organize The Chaos.*

THE ACHIEVER CYCLE

THE ACHIEVER MODE prides itself on its linearity. It gets things done. In a matter of weeks, months, years the achiever mind creates highways, bridges rivers, builds hydroelectric stations. A young person enters college through one door and several years later emerges from another as an accredited doctor or engineer or MBA. What went on inside the school to transform this person as raw material into finished product was a series of 'achievements'—goals set by the institution and realized in a step by step fashion by the individual being processed. Thus, cityman looks about him and sees things happening through the application of achiever energies and ideas to raw material. When time is conceived as linear and when one pursues goals considered worthwhile by society, there appear to be rewards. Much as the hunter in the earliest cultures saw the return of the salmon and the nuts and fruits of the uncultivated forests as his 'increases of nature', and the cultivator saw the harvest of emmer and lambs as his increase of nature, so the cityman sees his goals pursued and won as his own increase of nature. He believes that life has left no other way open for him to experience the largesse and significance of living. Without achievements and moving on the linear track, his years would be meaningless. There would be no reward. There would be no 'step-up' to the better tomorrow. Without achievements the achiever would be trapped in the 'chaos' of today.

The mind of cityman arises and exists as a tension between 'order' and 'disorder.' The result is a condition of continual becoming: a lifelong striving for 'what ought to be.' His mind lives, as it were, at the juncture between an unsatisfactory yesterday and a 'better tomorrow.'

Cityman will strongly assert that his thinking is *linear* and never approaches anything so abhorrent as going in cycles or circles! Cityman must set for himself—many times a day if he would be a proper achiever—the linear problem of moving from point A to point B by the shortest route. Yet, in what seems so simple a task the achiever has already deceived himself.

If one is at A in one condition and then two hours later arrives at B in a different condition, the linear argument seems to have been demonstrated beyond further need. In fact, nothing in this

A to B model appears readily contestable. The very words and grammar of the English we use are so biased in favor of linear movement and change as to render our investigation impotent before we are fairly started.

Consider, if you will, that when the achiever is in reality at A, his mind is not. That which comprises his genuine sense-of-reality is already on its way to B. But by the time he arrives at B, his mind, much like an old milkwagon horse that knows not to dally, is already on the move towards C. Man's mind under citythink adapts early on to this citygame of alienation (self at A, mind-reality at B). The mind, being another of nature's cyclic marvels, has in the city achiever mode no way of stopping and standing at B. This point is hardly more than a station that flashes by in the night. Now, under some other mode there might be a chance of pausing at the original A. In experiencing the mystery of A thoroughly, we might find that understandings of B, C, D, and so on were also included. Would not such a global insight far outweigh a mere linear shifting of images?

A casual glance at any one stage of the achiever mind in its image processing can give an illusion of linearism. But when a number of these movements are viewed together and linked as one ongoing process, they will be seen as so many stages of a cycle which, like all cycles, will in time curve back upon itself.

When at A the achiever feels himself in real time and devoid of purpose. Without a goal filling his mind and giving structure to his thoughts, he experiences a condition of unrest, uncertainty, doubt, anxiety. Without a specific goal to organize his time, he will flop about like a landed trout. Only when the achiever settles on another goal and actually leaves point A for point B, will he again become something approximating a whole person. As he attempts to find his way through difficulties to the new goal, most of his old capacities and self-assurance as an achiever come back into focus again. And he will declare himself 'saved' and a 'new man.'

On approaching B, however, the hero of tomorrow begins to feel mixed emotions. The first is elation. He has at last made it. But intertwined with elation—almost the back side of it—is a renewed doubt. Is this achievement really what he had visualized? Has the fashion for this goal been overrated or gone out of style? Does the attractive goal bring with it a host of threatening collaterals?

Almost invariably, the goal, which the achiever has worked so hard for and must now accept and deal with, turns into something that he had not quite expected. The goal in mind is like a mail-order bride (or groom) who does not match the photograph sent months or years before. Lewis Carroll gave us this amusing picture of an effort to structure a mutable environment:

> Alice thought she had never seen such a curious croquet-ground in her life: it was all ridges and furrows: the croquet balls were live hedgehogs, and the mallets live flamingoes, and the soldiers had to double themselves up and stand on their hands and feet, to make the arches.
>
> The chief difficulty Alice found at first was in managing her flamingo: she succeeded in getting its body tucked away, comfortably enough, under her arm, with its legs hanging down, but generally, just as she had got its neck straightened out, and was going to give the hedgehog a blow with its head, it *would* twist itself round and look up in her face, with such a puzzled expression that she could not help bursting out laughing....
>
> "I don't think they play at all fairly," Alice began, in rather a complaining tone, "and they all quarrel so dreadfully one ca'n't hear oneself speak—and they don't seem to have any rules in particular: at least, if there are, nobody attends to them—and you've no idea how confusing it is all the things being alive: for instance, there's the arch I've got to go through next walking about at the other end of the ground—and I should have croqueted the Queen's hedgehog just now, only it ran away when it saw mine coming!"[1]

The difference between the goal in mind and the achievement in hand produces a coefficient of incongruity that must be dealt with, not several times a day, but minute to minute. The question continually before the achiever is how much of less can he settle for. If an achiever does seem to arrive at his goal blithely and without incident, one may rightly suspect that, somewhere along the way, time's plaything has secretly swallowed, or still has to swallow, an amount of bile.

The achiever arriving at B and faced with the inevitable coefficient of incongruity has several options open to him. He can gamely accept his goal and try to live with it. To do so, however, will grind down his sense of being the capable achiever. Or, he can talk of the conflict between opposites and the subsequent dialectic compromise. He might also confess his secret strategy of setting goals high so that when he does have to 'adjust to reality,' he is somewhere close to what he wanted. The final choice in this dilemma of arriving at the goal is, of course, to discard the goal, to trash it, to pawn it off on an acquaintance. The roué tries to interest his friends in the mistress of whom he has tired. She no longer 'amuses him.'

Sooner or later, the achiever will dispose of the goal and look around for another. The stable situation for the achiever mind, we remember, is not standing at stage center, clutching his prize goal to his breast and affecting a silly grin of "I did it!" The climax of goal-in-hand is but one short step from completing the cycle and returning to anxiety. Having realized a goal also means being without a goal—being devoid of a plan for the future. Soon after the achiever has arrived at B and picked up his 'golden' ring, he finds himself, through a strange process of reversal, suddenly back at A minus his prize. In the interval from A to B he may seem to have traversed time and space. But in his mind he has gone nowhere. He is as before, always at A. No matter what actions he takes, he is ever the same modicum of uncertainty seeking emancipation. After each rebuff on this ladder of 'progress,' whose rungs keep breaking, the achiever loses a little more of his youth. And he grows increasingly distraught as he compares his situation with what appear to be the 'successes' of others.

The stirrings of the mind energies, no matter in what mode, recreate predictable wave forms and periods. The apologists of citythink persist, however, in sneering at earlier minds, referring to them as cyclic dead ends, bypassed and abandoned by evolution and progress. The hunter mind, we are told, was unable to differentiate itself from its external environment and associated its processes with recurring animal powers, or totems. And the cultivator mind was forever seeking to house itself in a panoply of ancient archetypes returned.

The achiever mode of the mind has a disconcerting, even mocking way of twisting about. The goal-oriented mind, though touted as linear, is cyclic. And, how could it be otherwise? Creation is itself a vast series of spinnings, turnings, and vortices. The four stages of the achiever cycle are: anxiety; search for and determination of goal; progress towards goal; achievement of and trashing of goal (which reenergizes anxiety).

Modern man is now divorced—alienated—from his environment and his past. Only rarely do natural events touch him. He rarely is subject to the cycles of nature—as the farmer still is—and he can move and use the world in ways his mind perceives most profitable. Because of this all-important release from the matrix of nature, modern man believes he can individuate his personality as never before in history. Goalseeking seems demonstrably more powerful and productive of effects than the older cultures based on the recurrence of significant traditions and archetypes. The latter is now perceived as 'circular thinking' and its lingering occurrences a kind of recidivism, the psychic refuge of 'losers' and 'wimps.' As opposed to this eternal return, the achiever 'gets things done.' His linear mind works, he has been told, from one problem to its solution, then goes on to another problem and its solution, and so on towards the distant future of man's own choosing.

What the achiever fails to understand, however, is that his perception has been impaired. He can no longer examine his immediate situation. When he is at A, his attention is centered on B. As a result the achiever will 'know' more about B than A. His experience of A is aversion and what he sees about him is the raw material by which to arrive to B. The result—and it is most significant—is that the civilized mind does not and cannot produce a clear, finely detailed picture of what is going on at A. The world and the immediate environment of cityman exist for him as a hazy, out-of-focus puppet show. What happens on the TV tube, however, he sees sharply, for those images take on an idealized reality. They are tomorrow—that is, "B" or "C" or "D"—made flesh through electronic magic. And they have a greater impact on his thought processes than what actually impinges on him. Without a maximum sensing of the nature and events of his immediate environment, such a 'spaced-out' individual tends to be gullible, compliant, and nose-led. To ask truly penetrating questions, to root beneath bureaucratic replies, to determine the nature of what is happening here and now, to be attracted by chance anomalies—these are not within the capacity of the upwardly mobile achiever. Early on, his mind was severely damaged.

❑❑❑❑

What would the world be, once bereft
Of wet and wildness? Let them be left,
O let them be left, wildness and wet;
Long live the weeds and the wilderness yet.

— Gerard Manley Hopkins

We look at it [Tao] and do not see it;
Its name is The Invisible.
We listen to it and do not hear it;
Its name is The Inaudible.
We touch it and do not find it;
Its name is The Subtle [without form].

— *The Way of Lao Tzu*

I am come as Time....

— Sri Krishna,
Bahagavad-Gita, XI

Time is the ridgepole of man's mind.

— C. C.

One must still have Chaos in oneself to be able to give birth to a dancing star.

— Nietzche

...seek within us what we can never find elsewhere, that we may find within us what no words can put there.

— Coleridge

Opening Hidden Frontiers

THE POWER OF THE HUMAN MIND, spectacular to witness, has always been the fearless march of a few into the unknown and their return with treasures. The problem—and opportunity—of dealing with the not-known confronts every individual. Each man, whether he will or no, is his own frontier. And on the edge of wilderness, time is always pinned at the now. For the wonders he comes upon, the individual to live in truth must coin his own words. Indeed, to *live* is to reexperience and to reinvent the world.

The battle in the urban world to command resources—and man's mind is a market resource—has ever been a war against the radically new. Cityculture's standing order is that man as citizen shall be discouraged from resorting to what is most native to him, the will and passion to inquire and the inventiveness to describe what he alone has found. Cityculture claims a patent on all innovations. To be accepted into the culture as lawful, each discovery must first be adapted and sanitized. When at last it appears publicly, it will have been institutionalized. Random innovations and runaway 'fads' border on the anarchic: they suggest revolutionary energies and uncontrolled vectors out of the underground. The innovative needs be quarantined lest the future be destabilized.

Whatever man knows about himself and his world is generally a replay of what his culture has seen fit to tell him. The very words we use to *think with* are property of the culture and deal only with yesterday and tomorrow. Our vocabulary and grammar channel our efforts to reconstruct the past or to plot forward in time for a 'better tomorrow.' But for expressions to describe that condition of mind which is time without the appearance of movement, we are deficient. Our culture cannot teach us the techniques and the psychology by which we might relate with the unrecognizable. For his experiencing of this most critical area—namely, what he does not know—the individual is totally unprepared and remains vulnerable throughout his life. At the age of twenty-one he seems more ignorant of his role in creation than he was at twenty-one weeks from creation's womb.

In every mind exists a frontier. At this edge, we may feel, but not recognize. We may sense, but have no words or concepts with the capacity to particularize. We may experience wonderments and fascinations, but find no help in our libraries or universities. The turning, at last, of the mind—from its long immurement in cityculture and the packaged future—back to the unknown, but fertile, present we are describing as the *discovery mind.* This is the non-metaphorical mind that deals directly with anomalies, strangenesses, and the concrete. The hunter permitted this mind naturally: he was born to wonder. Cityman, however, cannot now make a conscious return to the hunter-shaman mind. But when he does experience a moment of the frontier still alive in himself, the mind that responds to the unfamiliar is not quite the mind of the peasant, nor of the achiever, nor even of the hunter. It is a '4th mode.' Such an involvement with the not-known, at once alarming, joyous, and turbulent, is we shall call the *wilderness experience.*

Being concerned with the not-known, and even seeking the unmanifest, may in these times be interpreted as a loss of contact with 'reality.' Our education has worked diligently to make us familiar with the world. In no direction are we to look and find 'blemishes' of the unfamiliar. If something strange should happen to interrupt our well-plotted advance into the future, our reaction to avoid tangling with it will range from almost blindness to outrage. We are to sidestep adroitly as one might a fallen soldier and charge on. And if the intrusion insists on delaying us, we can either dispatch it or protest that our 'privacy' is being intruded upon. We live in the information age. As 'responsible adults' we feel our lives are already crowded with more input than we can possibly handle and must discriminate. So, what chance does some interruption, some anomaly have on our time? The child's urge to question is a luxury we can no longer afford. When an oddity jumps out of the bushes, we may say, "I'll have to look that up." But having gone that far, we can forget it. Explanations are for books and we have libraries full of them. To make an issue of "facing the not-known" suggests to citymind a kind of nihilism, a denial of rational thought, a rejection of all that 'civilization' has accumulated. If you challenge cityman's reliance on familiarity, he will probably counter with, "Why reinvent the wheel?"

In all ages, however, some men have turned from their culture to the unknown. The hunter-gatherer—whether the ancient bison hunters of the Lascaux Caves or, twenty-thousand years later, the bison hunters of the Montana prairies—accepted adventuring in the unfamiliar as a normal, if astonishing, happening. Not everyone had the second sight to 'sense' where the bison were wintering, or where to locate lost children, or what danger threatened the encampment. Nor did all experience the same magnitude of power visions. But those less gifted were never without a taste of these experiences. They understood this dimension of the mind of which we now are so ignorant. It was an accepted fact that some appeared more adept than others. And to these shamans honor and attention were paid.

The peasant-cultivator who 'displaced' the hunter was well-acquainted with stories of earlier men who journeyed beyond the familiar and returned with strange knowledge. And of such persons they were thoroughly frightened, depicting the shamans and dancers of god as 'evil' and 'devils.' Similarly, the men who appeared out of the wilderness to stand at the gate of the agricultural village and professed to 'speak the word of the Lord' were not always well received. Some were stoned rather than heeded. Men who would supplant the familiar with the unfamiliar, who would change the meaning of time, are forever suspect. If not looked down on as eccentric, or crank, they are feared as upstart, disruptive, unnatural.

LIVE THE FOREST

THE TRADITION-PRESERVING MIND of the planter and the destiny-changing mind of the kingship united in wiping out the hunter. But the fierce running battles they fought between themselves would, from then on, seem the main plot in man's official history. Hidden by this well-documented conflict, the older psychic stirrings continued on. The discovery mind—the unconditioned, or nonlinear, which looks neither to yesterday nor tomorrow—would also have its play. In the time of the hunter the mind focused upon the unfamiliar was not thought of as an aberration, but was supported. And regular attempts were made in those dawn ages to encourage its appearance and its transforming experiences. Despite its persistent, if shadowy, history since the loss of the hunter, the 'discovery' mind has yet to dominate an entire epoch as did the peasant-tradition mind and as presently does the kingship-tomorrow mind.

The discovery mind, though infrequent of appearance, is nevertheless the wild card, the joker, that can change the whole game. From time to time it has popped up out of nowhere and created outrageously dramatic scenes. Its appearance might be likened to a summer storm in the higher mountains. What seems a pleasant rain is soon followed by nearby streaks of lightning which fairly singe your hair. At the outset, novelties of storm and of insight may seem entertaining and welcome. But as white flashes shatter nearby trees and set them on fire, as new insights demolish old verities, such manifestations of power can be frightening. The appearance of even one discovery man can upset and redirect other men's thinking for centuries. "The world," Nietzsche wrote, "revolves—invisibly revolves—about the inventor of new values."[1]

The early peasants stood in awe and no little fear of the 'abnormality' of the discovery mind which 'saw' what others did not. Many perceived it as an epiphany—an expression of divinity—and the prophet-shaman as a divine madman. The books of 1 Kings and 2 Kings contain many tales of this special man and his supernormal powers. To traffic with the not-known was to speak with the gods, to fly with the winds, to be able to 'call down fire from heaven.' Every kingship fears subversion of its control of time. It abhors most the disruptions inherent in the outbreak of 4th-mode skepticism and innovation. Every kingship—early or late, personal or modern institutional—feels compelled to capture, isolate, and channel the discovery phenomenon, no matter the cost or means. Great efforts were made, we know now from the obliteration of the Jerusalem church and the Gnostics, to modify and monopolize the Jesus story for the new masters of time.

Only in the last century have we been able to recover much of that history which the Bronze Age and Iron Age kingships took pains to burn and cover with the sands of the Ancient Near East. The rapid acceptance of the Jesus story throughout the 'known world' was largely due, it now seems clear, to accommodation. All during antiquity the most powerful myth by far was that of the divine mother and her son. He it was who offered himself as the yearly spring sacrifice which would assure the perpetuation of the agricultural cycles—and thereby 'save' mankind. In one age the 'queen of heaven' was Ishtar and the living-dying god was Tammuz. In another, they were Isis and Osiris. In still another, Aphrodite/Persephone and Adonis. When the missionaries to the gentiles apotheosized Jesus to the role of Tammuz-Adonis, his acceptance by the larger world, thoroughly accustomed to this agricultural-savior-god paradigm, was greatly eased. (Today we might say he had been given a look-alike label.)

The impact of the wilderness Jesus, however, the shaman reincarnating the unpredictable powers of Elisha and Elijah, was steadily downplayed and finally denatured altogether. Certain remnants, however, of the original Jesus and his role as iconoclast and 'time-changer' remained as worrisome anomalies embedded in the Gospels. Passages in the Sermon on the Mount which begin with such wilderness-turning phrases as "Consider the lilies of the field" still give 'right-minded' ministers who promote Jesus as the young god of tomorrow much trouble—as we shall see.

When we inquire into the non-metaphorical, or 4th mode of the mind, we leave the realm of words and the rule of institutionalized cultures. We enter into the forest of our own wilderness where time and the changeless would work their metamorphic ways on our bodies and minds. Cityculture would have us believe, however, that we are departing reality for the psychotic and vicious: where 'black arts,' voodoo, incantations, the ecstatic, the occult, and the unscientific hold sway. We will be adrift in the world of spiritualism, trances, and fortune-telling, the realm of the fevered brain where visions, voices, and the fantastic take over and destroy man's hold on the real and solid world.

Beware of your mind! we are warned.[2] Avoid it as you would a 'black hole.' When you feel yourself sliding out of thoughts of tomorrow into the morass of today, turn up your Walkman. Or, plug in a video cassette. True enough, the 'general issue mind' which we know as citymind is a condition scarcely tenable. With ricocheting desires, competing images, and bat-like aversions flying in every direction, citymind *is* a no-man's land. The achiever mind—as supplied by the city and founded on conflict and tomorrow—now that surely is a mind to be cautious of.

The mind of wilderness, on the other hand, we are scarcely aware of: the psyche which we left far behind in our childhood was a faculty never allowed to develop to its own adulthood. It remains a vestige, like an arm shriveled by an early attack of polio.

The shattered thinking and phantasmagorias we experience with the citymind are hardly organic. These images of violence and of desire which flood and disrupt our psyche are disorderly, but *recognizable*. And here, we must remember that whatever is possessed of familiarity, likeable or not, belongs in one way or another to the culture. Man's natural mind, on the other hand, is unconcerned with the familiar. The product of that mind is the experience of wilderness. And what the mind encounters on the frontier of the not-known is beyond recognition, well beyond the range of the culture and its vocabulary and grammar which establish the familiar. The wilderness encounter is a wordless experience. When it happens, it comes as an image-vision or as a thrilling hunch or as a "queer feeling." No words rise easily to encompass the experience of living the forest.

That world which once acknowledged the prophet, the seer, the oracle, the intuitive clairvoyant, and the divine 'madman' is now a 'lost world.' The very presence of the individual of larger vision had been a comfort. He was proof that creation, for all of its mystery, was accessible and that man was both its kinsman and intimate expression. The onslaught of cityculture, however, has now swept away the last unaffected examples of the early hunter and the early planter. Hundreds of thousands of years of experiment, learning, wisdom, and gifted interaction with creation have been destroyed. In addition, those expressions of direct insight which the kindred people treasured as gifts and visitations from the spirit, or power side of life, were discredited as devilish, infantile, or were burned altogether. Today, when our fulfillment is exterior and vulnerable we find it difficult to understand a people who, if simple outwardly, might be inwardly complex and rich.

The persistence among us of this mind which now and then seems capable of 'second sight' and of strange prescient 'guesses' is the survival of a faculty once fairly common. Among some peoples it is still highly treasured. The disaffected of all citycultures have looked to the survivors of 'less-developed' cultures for religious inspiration. The Egyptians, to the little dancers of god from Central Africa; the Greeks, to the Scythian wizards; the Romans, to the Jewish-Christian diaspora; the Hellenic remnants, to nomads out of the Arabic hinterlands; the English and Americans colonialists, to mystics out of Indic caves. Gurdjieff disappeared for a period of years and sought dervishes and other holy men in the remote mountains of Persia, Afghanistan, and Tibet. Mircea Eliade went east to become the disciple of an Indian guru. In the 1970s young Americans searched high desert reservations for latter-day Black Elks. In all ages citymen have hoped to find in the 'pre-civilized' what they themselves had not—integrity, strength, and insight. They looked for men who might still be living in the way of the evolutionary mind. For the power to detect the unexpected and to experience enlargement of understanding arises from the natural world and not from the ersatz and manufactured.

EARLY AND LATE, all cultures have found one problem more difficult and crucial than all others. To avoid dealing with it has often proved their undoing. That special problem is how the radically new shall be detected and 'safely' introduced into the mainstream of the individual and collective life.

Older agricultural societies and early kingships generally resorted to simple avoidance and, if necessary, repression. Our latter-day institutional kingship steadfastly champions 'constructive change,' but reserves the right to be the channel by which the 'new' shall be recognized and introduced. This is simply avoidance of another sort. Cosmetic surface changes may be allowed, but no shifts of the approved, underlying paradigms. The agenda for what will be considered news is set for the whole country by a few newspaper and media giants. All others copy. The key purpose of the media is to occupy that niche between yesterday and tomorrow so effectively that today's passing will scarcely be noticed. Thus we are surfeited with every kind of kill-time busyness: happy-talk journalism, eye-popping spectacles, sexual divertissements, pageants of 'historical importance,' celebrations of movie heroes, and staged debates on the 'issues' of the moment. What would focus our wordless attention on a live today—as did those inadvertent clips of torching Vietnamese villages—appears rarely. Not made visible by the media of record, that which should be disruptive may never ruffle the stream of history. Ergo, it will be doubtful such an event ever happened. If you do not acknowledge something, what existence, what effect on man's culture can it have?

The perennial question, once more, is how the introduction of the radically new, the revolutionary new, the challenging and unexpected new shall be handled? We go from day to day believing we have time under control. Events on our calendars shall proceed in a mannerly fashion, according to either traditions out of the past or scenarios for the future. But suddenly, along comes an uncalled-for, unaccounted-for 'happening'—a turbulence which, by trashing our calendar, challenges our right to choose. The new factor could be a baby, a change of the tax law, a foreclosure, the illness of a relative, or even a wholly new perception of the thrust of one's life.

Between the individual and the culture that tells him who he is (formats his mind) there exists a peculiar symbiotic relationship. It so happens that of the two only the individual inherits the faculties which can sense the nonverbal impressions by which the radically new communicates its presence. The culture, on the other hand, is wholly verbal, wholly committed to the remembered past and to the engineered future. But to the present the culture is hostile, and understandably so. For in that area of time remembering and engineering are of no use. The present can only be perceived in a wordless manner—that is, as a whole, living image. Depending solely on language and its manipulation, the administrators and apologists of the culture are inhibited from participating in today. Both the culture and those who are its immediate creatures lack, therefore, that which is vital if hints from the unperceived are to be detected. Their reactions are always expressions of contempt and anger towards the unscheduled and contrary.

Man's mind is so organized, however, that it can deal directly on nonverbal levels with the unexpected and the radically new. Such a faculty has been much discouraged early and late. The discovery man restructures, almost by his presence alone, some important part of the known world. What had seemed settled and perfectly ordinary becomes through him mysterious, unknown, and even disturbing. Never again can it be perceived in the old ways.

The discovery man brings wilderness and 'chaos' into the midst of the settled camp or city. He may be a shepherd worrying himself sick about his people who are slaves in Pharaonic Egypt. Or, he may be a hairy man wearing a camel's hair shirt and a wide leather girdle who stands at the gate of his besieged city, anguished at his community's loss of cohesion and kinship ways. At the passing donkey riders, charioteers, and people on foot he thunders out the strange words and prophecies which he is 'hearing' with his inner ear. Or, he may be a Mediterranean gentleman lowering himself into his bath, thinking about the special relationship between the weight and mass of specific metals. All at once there flashes into his mind, as we have been told, the perception of what he had not thought of before, nor had any other man. And he leaps out of the water and dashes into the streets, shouting, "I have found it! Eureka!" Or, he may be a young English mathematician, his university closed by the plague, who now has a bit of leisure to play with some old puzzles. While he rests in an orchard, his mind fatigued with contemplating the mysteries of heavenly bodies and their orbits, an apple falls to the ground in front of him. And just as suddenly, the perception he sought has found him. Still again, he may be a monk, who patiently tends a small plot of peas. Something mysterious about the subtle differences among his legumes catches his attention. With consummate patience he follows his wonder and allows the pea vines to teach him some of the wisdom of natural variation. Then again, he could be a grandfatherly mid-European, puffing on his pipe and looking at us with his sad eyes, who tells us of what he has found, the interchangeability of energy and matter and the relativity of space and time—his formulas becoming the first to quantify the powers locked in the atom.

What distinguishes the discovery mind and its tour into wilderness is not just its intention to exceed what the average or even exceptional individual knows. This mind would range outside whatever has been incorporated into the wisdom of the culture itself. As he matures, the man of wilderness may often grow disdainful of the body of knowledge that is termed education and those who would presume to 'teach' him. For the culture's novelties, improvements, 'advances,' and other glitter he has little patience. His concern with the unknown is not with established areas of study he knows he does not know. Nor is it with what most men do not know. He would instead go beyond, to that which he does not know he does not know. He would enter the unthinkable, unwordable, vertiginous regions which frighten all men. He would investigate what has never been found in books or even on the tongues of men.

This pursuit of what appears irrational—outside the cultural norms—is not easy. The individual on the wilderness path is out on the edge of the world with no visible help. The culture conceives of itself as complete and self-fulfilling. Nothing exterior to the culture is needed for a rewarding life. If something extraordinary, however, should be discovered, it will be attacked if it appears to question the basic time assumptions of the culture. Otherwise, it will be ignored or denied until it can be rediscovered from within the culture. Arthur Koestler wrote of the pathology of scientific thought:

> Religious and political oppression play only an incidental part in the history of science; its erratic course and recurrent crises are caused by internal factors.
>
> One of the conspicuous handicaps is the conservatism of the scientific mind in its corporate aspect. The collective matrix of a science at a given time is determined by a kind of establishment, which includes universities, learned societies, and, more recently, the editorial offices of technical journals. Like other establishments, they are consciously or unconsciously bent on preserving the *status quo*—partly because unorthodox innovations are a threat to their authority, but also because of the deeper fear that their laboriously erected intellectual edifice might collapse under the impact.[3]

Discovery and invention, as recognizable, cityculture pursuits, have been institutionalized. At the turn of the century, with the real frontier becoming only a memory, many still sought a semblance of the frontier wherever they could. Some retreated into barns, closed the doors, and built airplanes and motorcars. Airships were stitched out of muslin and filled with illuminating gas. The source of the Nile and the position of the North Pole became vital matters. Such questing by individuals for the unrealized, the unachieved—much less genuine discovery of what has never been previsualized at all—is by now quite rare. An outbreak of frontier fever occurred recently in a nearby community that became known as Silicon Valley. But even there, the wilderness was soon mapped and subdivided.

Everywhere the role of innovator is rapidly co-opted by our institutionalized kingships—the so-called private and public sectors. The very meaning of discovery is thereby changed in character. No longer will the unknown be permitted the chance to expand, much less alter, the consciousness of the discoverer. In the cityworld discovery becomes little more than a means of extending kingship control. Anything worth discovering must be usable. And anything usable, it is declared, will be discovered by cityculture scientists. If they are not already working on the subject, the 'problem' does not exist or is not worth investigating.

The mind-set of our would-be investigators has been further distorted by a fixation with grants. Upon locating a point of entry into the undiscovered, most of us will move no further. We must write up proposals to Guggenheim, the National Endowment for the Humanities, or some other institutionalized philanthropy and wait for a subsidy. Our venture into the unknown is placed at risk before it commences; it must gain the attention of bemused strangers and compete with other projects on the basis of its ultimate usefulness. If aid from one of these cultural regulators is not forthcoming, our effort to experience our own discovery mind shrivels.

When the man of passion senses the proximity of the unfamiliar, he will dive into that unknown without delay and however he can. He could not do otherwise. To him the approval of strangers is meaningless. Help from the outside only tends to blur his sensibilities. Further, how can he describe convincingly that which he himself has not grasped?

The creative pragmatism of modern cityculture is not your true wilderness experience. The former is a novel accommodation to given paradigms. In the wilderness experience the practicality of the discovery itself is secondary. The major event is the expansion of the individual's mind and his changed relationship with his world and himself. A considerable revaluation will have taken place: the old constraints and problems no longer look the same. Where mere use-factors of the discovery are held paramount, however, such wilderness venturing is the citymind in another guise. If you must look before you leap, if you must wait for assistance, then your sighting of a frontier did not presage a true wilderness experience after all, merely the old colonial achiever scenario all over again.

WE SWIM IN MYSTERY. We cannot avoid it. There is far more unknown within ourselves and all around us than is known. Mystery, or the not-known, is the only dependable reality. What we think of as the familiar is suspect by its very nature and must be thought of as tentative. Being cast in words, the familiar is at once both the creature and the snare of the culture. Again, each word is an image which would freeze time, either into the past or into the future, and defeat the uncertainty of the instant. The familiar word enables the accultur-ated mind to project onto the unmanifest what it expects to see, hear, taste, smell, feel, or otherwise perceive. The organization of the achiever mind simply does not allow it to obtain a detailed, penetrating picture of the immediate. Whatever it thinks it sees and hears is blurred by aberrations produced through expectation. Being so limited, we are like a myopic driver without his glasses. In a world we are not sure of, we should be moving with caution. The familiar ought to be approached as an uncertainty, if not an outright illusion.

Those minds that become fascinated by an improbable directly at hand—where others see only the ordinary—are exceptional in any age. Some of us might be attracted for a moment, but the mind which will hold on, even when the mystery with its difficulties would shake him off, now seems rare. For the mind in this state, or mode, exhibits a persistency past any reasonable hope of result, of answer, of help, or almost of a miracle. Such a stubbornness, which we also find in Job, will appear to others as a disease and might be compared with Ahab's pursuit of the white whale and the death-grip of an English bulldog. In the course of this fist-to-heaven defiance, however, ordinary reality for the discovery man begins to melt, familiar shapes becoming less certain. The simple and explained now become the complex and unexplained. What was thought real is now seen as unreal. Only the mystery which continues to draw his attention and which refuses to be encompassed by words is the genuine.

It is in this moment of 'breakdown' that the wilderness experience may occur. The old reality has fallen apart. The field now consists of many components each asserting its independence. After accepting this confusion and living for moments or for years in unfamiliarity, new, unforeseen relationships between the components may at last emerge. And the alterations produced in the individual's feeling for himself and the world by these accesses of intelligence can be profound and life-altering. An early shaman might describe his wilderness initiation as having had his bones stripped of flesh and reassembled.[4] The fresh, expanded perceptions redefine time and greatly alter perceptions of what matters and the orders of significance. This revelatory experience becomes a central reality for the individual and its felt intensity will remain for a long while. Under propitious circumstances it may be transmitted to and incorporated by his community.

This moment of revaluation and perspective-shifting is generally experienced as an epiphany. When the 'new' bursts upon the receptive mind, the individual may feel—and, indeed, why should he not?—that creation itself paused and 'spoke' directly to him. It took time out, as it were, and revealed to him, and to him alone, some special aspect of the universe which no one had before suspected. The moment of insight has—as we shall see in the forthcoming essay, *Life Of The Mind*—been variously described. It is a surge of elation and wonder. It is also a humbling sense of having been touched and authenticated by the powers of creation. The metaphor of illumination and a divine light has been much used. Shamans also spoke of their organs being replaced with potent, light-emitting gemstones.[5] Many of us still describe an enlargement of perception as a "dawning."

The discovery-mind phenomenon creates in each appearance a strikingly unique person. Underlying all such occurrences, the mode, however, is unvarying. The trance-coma of the shaman; the 'absentmindedness' of the scientist; the chanting of the enraptured poet; the austerities of the mystic; the inspired vertigo of the whirling dervish; the wilderness-wandering and fasting of the prophet—all these have much in common. Each in its own way is confronting the not-known. And we shall be particularly interested, as we proceed, to note the various stages in the unfolding and development of the discovery expression and to mark the basic similarities in certain exemplary intuitive, or "4th mode" expressions. In every case the discovery man, once he has been attracted by his special mystery, becomes absorbed in it. He plunges like a setter into a marsh after a distant bird seen falling from the sky.

The mind of discovery is not some peculiar psychic extrusion or an unnatural mutation of the mind. One does not have to hold one's breath or bend one's self into a strange shape to arrive at this 4th mode. Rather, the state of mind in which we now find ourselves, and call reality, is a considerable distortion. The wilderness-facing mode is actually man's mind recovering its natural state: its emergence represents a 'fall back' into a stable condition. Here, we come upon a special problem. However much he should be intrigued with its presence and power, the individual cannot reach and activate what we are calling the 4th mode. His will to achieve is here of no service. Efforts to propel his mind into wilderness—to have a taste of the unfamiliar—will only confuse him. He will find that the epiphany—or that expansion of consciousness when the unfamiliar reveals itself to him—will ever elude him. Only following a breakdown, small or large, of his current thought mode will there be any chance of that marvel occurring. But so unnerving can this lacuna of the familiar be that he will more often flee the adventure rather than hold it to his bosom as his own jewel. Until then, his potential for the wilderness experience will remain hidden. It may surface now and then, however, as part of his dream world. The nightmare is often a nudge by the nonlinear mind to recall us to our unrealized heritage.

Those acts of discovery and revelations of the new which change man's relationships with the world and with himself are invariably associated either with Divinity or with a quasi-divinity—such as the angel Lucifer. Thus, the emergence of what was never before revealed—the insight, the revelation, the vision, the epiphany, the oracle, the moment of satori, of nirvana—bespeaks the activity of the godhead. The recognizable is the human world; the unfamiliar and unrecognizable must, therefore, come from another realm. A divinity we define as the controller of all unknowns and their potentials for good and ill. Living exclusively as we do in the familiar (or alienated time), we have come to believe our encounters with the unknown are somehow either rewards or punishments.

The unconscious, ongoing need we have for a god is to have access to a master of time. Cityman is not in awe of creation or the wonder of his being here. That is an accident as far as he is concerned. What he wants is a defense against the unfamiliar and the unexplained which pop up on his way to tomorrow. We feel at greatest risk in the uncertainty of the present. We seek not the godhead for itself, but for a dispensation of its special power with which to break the willfulness of today. It is out of that resource we would fashion our separate arenas of time as manifest and predictable—either the paradigm rewards out of yesterday or the glittering promises out of tomorrow.

QUARANTINE OF WILDERNESS

IN OUR CULTURAL HISTORIES one can find various phenomena described again and again as special acts of God. Thunder, for instance, was at one time God's voice and lightning was His bolt. Women conceived because God opened their wombs. Early diviners, seeking to read the contours of time believed implicitly that God would show his Hand by influencing the roll of the sacred die, the spin of the Urim and Thummim, and the fall of the I Ching sticks.[1]

The term *God* has always been our metaphor (our algebraic symbol, or "X" if you will) for the unexpected, for the unplanned, for that which lies (or once lay) beyond human control and human understanding. The movement of the unknown—of whatever is going on beyond our ability to recognize—becomes by this rationale the movement of the Divine. And to reveal the nature of even a small aspect of this enormous and puzzling unknown may itself appear superhuman.

No more are we permitted the conceit of thunder as divine anger or the moon as the habitation of a goddess. The slender, heaven-pointing rods invented by our homespun Yankee wizard defeated the mighty hammer of Thor. Still, that apocryphal time-god known to us as *luck* is regularly invoked to guide our endeavors. To be sure this is a term we no longer consciously associate with a god and with divine power over time. But in our hopes and prayers for "luck" to control future events, that which we are invoking serves every bit in the capacity of a god for us. The O.E.D. does not give the Norse trickster god *Loki* as the source of 'luck.' Yet, the cognate sounds and meanings in Loki-luck are marvelously suggestive.

Early man as peasant, living with what he perceived as exhibitions of sacral power, conducted himself with considerable caution. The Unmanifest, he believed, could be thought of as cosmic clouds of energies. At times, these 'precipitated' in a fairly orderly manner. The appearance and growth of plants was one such, fairly predictable infusion of 'holiness,' 'numen,' or 'mana.' Another was successful insemination of the females of the various domesticated species. Great care had to be taken, however, to avoid provoking the Unmanifest into unruly "outbreaks" of divine potencies. For then, havoc could be unloosed on the Earth. Storms, clouds of locusts, tidal waves were perceived as just such anarchic, unregulated, angry appearances of numen. In Melville's epic,

Moby Dick, we are held spellbound by not one, but two dangerous eruptions of anarchic holy power. The one was the Faustian sea wizard, Ahab; the other, the power-mystery of the ocean, the White Whale. The tale of Pandora's Box was no children's fairy story, but a stern warning of unseen potencies which could break forth in the lives of the unwise and hubristic like a great storm. The anxieties which the peasant felt over contamination by an excess of the "holy" have modern parallels. Today, we are very disturbed by "outbreaks" of the unfamiliar and dangerous: we are faced with contamination of our water and soils with toxins; our tomorrows have become hostage to crushing debts and nuclear threats.

In the early agricultural societies the harvest ceremony was not just to give thanks. Its primary purpose was to *desacralize* the grains and fruits. Otherwise they would be dangerous and could not be consumed. The busk, or green corn ceremony of the Southeastern American Indians was necessary to neutralize the divine potency still radiant and, therefore, lethal in the new corn.[2] In an earlier time the hunter had treasured his contacts with the unmanifest. In all planting cultures, however, the peasants were easily frightened by the unusual. They felt threatened by any slight shift of the traditional which might endanger their crops, their animals, and their own fertility. Wonders which seemed touched with excesses of holiness were scrupulously quarantined and separated from their sphere of familiarity for fear of disruption. The woman's menses and the newly dead were often subjects of much concern and strict ritual. The Ark of the Covenant was the very embodiment of the peasant's terror of the unpredictable. It was carried into battle to work its time magic over events on the field. (This strange mix of fear and holiness still works in our cityculture and has resurfaced in our perception of "the flag" as a quasi-holy embodiment of the nation.) When David would return the Ark to his city, there was an accident:

> David and all Israel danced for joy before God without restraint to the sound of singing, of harps and lutes, of tambourines, and cymbals and trumpets. But when they

> came to the threshing-floor of Kidon, the oxen stumbled, and Uzzah put out his hand to hold the Ark. The LORD was angry with Uzzah and struck him down because he had put out his hand to the Ark. So he died there before God. David was vexed because the LORD's anger had broken out upon Uzzah.... David was afraid of God that day and said, 'How can I harbour the Ark of God after this?'[3]

Like the Ark, like the green corn, the prophet out of wilderness was perceived as an expression of the unmanifest. He was, therefore, brimming with excesses of numen. A kind of divine voltage charged his body. Some might seek to touch his robe or "take the dust of his feet" in order to gather some of this blessing. Elijah was such a one who could call down "fire from heaven." The man of the timeless was possessed of such holiness, it was thought, that he could master space and time. He could see from afar through visions and the dead would rise at his touch. The mission of the messiah was to bring a 'new order.' One approached such a marvel, therefore, with the same caution and respect as one might the Ark or some other sacral object.

The peasants around the world developed various ceremonies for desacralizing their crops and "increases of nature." While the Cherokees were busy with their green corn ceremonies, the Semites and others of the Eastern Mediterranean devoted (read, "sacrificed") the first of their harvests and births to their fertility gods. Nor did they exclude from these rites their own first born: Abraham's compulsion to immolate Isaac on a "high place" was no anomaly. The command he 'heard' for offering first fruits was backed by traditions thousands of years old. To go contrary to such a powerful cultural imperative would indicate an historic redefinition of time. In Abraham we have the suggestion of such a shift. He was not one, however, to go all the way and say that the god of this "high place" could get along without any sacrifice at all. The first-fruits ceremony was still 'necessary' to contain outbreaks of the numinous and assure an orderly continuance of all manifest phenomena. According to the early legend, Abraham received heavenly permission for downgrading the sacrifice. Some bold Abraham, we can surmise, (or more likely, a tribe named Abraham) replaced the firstborn child with a ram. It was hoped the switch might go unnoticed and that the smoke of the latter would be as "pleasing and soothing" as smoke of the former might have been.

The peoples of citycultures, early and late, have in their turn experimented with ways to defuse random energies which might collect and grow dangerous. The man of wilderness, however, was one who brought a whirlwind in through the city gates. And such acultural, anarchic powers on the loose have always been the subject of concern. Explicit in discovery-man is an invalidation of the familiar. Implicit in him is validation of the 'ecstatic'—that is, every individual's potential to sense and experience what he has not before perceived. "Hunch," "insight," "epiphany," "ecstatic," all speak of a similar happening: the thrilling moment when the mind suddenly incorporates an enlarged understanding beyond anything the culture has provided or allowed. Of course, such contamination of the conventional time order had to be, and still has to be, handled without delay and in whatever way expedient, lest 'civilization' be turned upside down.

If the prophet's new ideas of time were appealing and if they caught on, whole populations might be destabilized. The cheapest way of dealing with such a redefiner of time has been to surround him with a curtain of invisibility. But if ignoring him will not work, impeach him and his rabble as terrorists. Stage confrontations. Blame them for the disturbances. Frame the leader with a heinous crime. Call upon the populace to cleanse themselves of this blotch. Repress his followers for several generations. Then gradually co-opt the disturbance. Show that the ideas introduced by the great man were cityculture's after all. When the outbreak of wilderness energies has been diffused, give the devil his due: raise him to sainthood. Then he becomes a safe talisman *for* cityculture.

Apotheosis creates an effective quarantine. By declaring the discovery mind an exception to the human race, the community functioning in citytime isolates the man of wilderness. It may thereby continue living as before, shielded from what would challenge its own peculiar meanings of time. The 'outbreak of holiness' always represents a situation of crisis when time-as-today threatens to overwhelm time-as-tradition and time-as-tomorrow.

After the outbreak of wilderness-holiness has been shunted into yesterday, whatever embers remain of the fire from heaven can be mopped up. Aroused and angry followers will accomplish little once their energies are diverted from today and channelled into hallowing the memory of their "madman of god." Even more energy will be drawn off if they hope for his return sometime in the future. Such areas of alien time are safe because they do not belong to the people. Besides, official deification (nowadays in the form of coins, stamps, holidays, and street names) will tidy up whatever hard feelings remain. In addition, it justifies quite handily the continuation of time as alienated.

Once our discovery-man has been safely encapsulated through apotheosis, his image may be confused in a calculated fashion with the standard myths of the epoch. (Jesus became a latter-day Adonis, Attis, Osiris, Tammuz.) His fame will increase thereby exponentially. And everywhere images of him will be kept on view behind railings. Viewers may approach these shrines and wonder at the god-man-anomaly like they might a meteor or 'black stone' fallen from the sky. He will be advertised as an outbreak of holiness that happened once and will never occur again.

Because of such distortions the ordinary individual will grasp little of the psychology behind this discovery experience. Instead, he is taught to fear the holy, the ecstatic, the unfamiliar, and his own mind. Yet, the man of wilderness, even at a great remove, exerts a considerable fascination. He seems authentic in the way that a mountain is undeniable and mysterious. Yet, what to make of this 'wrestling with God,' of this forty days in the Wilderness?

Are not these heroics something which only a god can undertake and undergo? Is not the man of wilderness our intermediary? Is he not a step-down transformer between the dangers of unmanifest holiness and our own material existence? He alone can touch the Ark—take the voltage—where we would surely perish. So effective are the inhibitions we accept that if "the Glory of God" were to present itself to us directly, we would not dare, as cityman, to behold the brilliance for ourselves. But we might kiss the hem of one whom we believed had seen the unseeable.

Yet all the while, cityman is as genetically driven to seek the unfamiliar as is the man of wilderness. His mind would naturally hover between yesterday and tomorrow in that narrow time-slot which is the very home of 'chaos' and the numinous. For here is the source of all that is fascinating and truly *alive.*

Man's natural psyche, then, is the discovery mode: namely, that attitude of mind which would authenticate us through direct contact in real time with the not-known and unrecognizable. But all this, cityculture steadily blocks by means of its powerful aversions. Such 'quarantines,' even though set in place centuries and millennia ago, still affect the fruits of man as before. We must now live with these painful inner conflicts. They inevitably arise in us because of our being cut off, through culturally induced fears and loathings, from access to our own natural wilderness authenticity.

In any age or society the discovery man is the source, not of the faddishly new, but of the revolutionary new. He is the explosion of insight. In him creation explains a little bit more of itself. The discovery man appears as our only personal contact with the creative urge of the universe. We search through his life and his words for clues as to why we are here, what we ought to be doing, and what creation itself might intend. In him we seem to catch a few glimmers of meaning amidst eons of chaos. Our relations with this wilderness phenomenon are, of course, a paradox. We destroy the man, but venerate his *memory.* In every way we shield ourselves from the rigors of our own direct experience with the mind in its timeless, discovery mode. And we prize the traditions of those occasions when it did happen to a few. For even the tradition is charged with numen. It gives us some feeling that reason, recognition, and caring do exist in an exploitive, alienated world.

The discovery man, then, is one who feels it is all right to be lost, to move amidst the shadows and the unrecognizables, to live the mystery. For him it is not unnatural to confront the strange and the unknown. He does not regret finding yesterday and tomorrow unreliable, even nonexistent. In fact, he has learned to relish the onset of a certain unsteady feeling—as though he were about to experience a fever or as though he were in a small boat caught in the waves of a large sea. This characteristic loss of balance, or 'about to fall' feeling, signals him that he has again chanced upon an area of the unmapped, that he is again venturing close to the hidden entrance to the not-known.[4]

In the discovery man there is the development to maturity of that which most lose early on. Man is an organism that comes equipped with what is needed to relate and deal with that which is beyond familiarity. The hard-edged, recognizable, tangible personality he is supposed to assume as a fully realized cityman achiever is an unnatural and binding construct. Even today there are individuals who have found that 'falling into chaos'—or, a state of not knowing—is an experience to be welcomed. For the man of wilderness the plunge into the undefined becomes more natural and vital than to cling desperately to that crumbling fiction of a familiar tomorrow which we call the real. Only out of relating to 'chaos' can an understanding approaching a true reality emerge. A lapse of recognition must, therefore, precede every new state of order, or perception. No revelation of the broader conception and intents of creation can emerge except through 'chaos.' But most of us shy from turbulence and feeling loss of control. We are schooled to prefer the comfort of recognizing the familiar. With our hand on the TV controller we search from channel to channel for a face we might know and identify with.

LIVE THE MYSTERY

IN ALL AGES WE FIND DISCOVERY MAN AT WORK on the problem of how to detect and incorporate the unfamiliar. To those of us functioning in the traditional or achiever modes, however, the encounters and scrapes we have with the strange and unfamiliar leave us feeling upset, nauseous, and even frightened. How can one be expected to relate with what is beyond the known or imaginable? Furthermore, by what means can one bring into the field of the known, where time is a precise gridwork, that which is beyond words, without category, and might, therefore, 'distort' time? To find some strangeness and drag its carcass across the cityculture's neat categories would be like fishing for trout and landing the Loch Ness monster! Finally, the peril of tilting with windmills has been restated in our day as, "You can't fight city hall." What is to be gained by the individual, we want to know, in this romantic pursuit of the uncertain and unfamiliar, but a string of difficulties? Why should we search in 'nowhere' for a 'new reality' which might crumble the foundation stones of our present and long-enduring order?

The conventional mind wants to proceed from a base anchored in the known and familiar. It would nibble daintily and only once in a while at the unknown. But creation does not play such a conditional game and it does not seem intent on us doing so either. All our familiarity schemes are sand castles, subject to periodic meltdowns. The natural world wants to draw us into games of total engagement—to risk the death of the conventional mind and to invite the fever of not-knowing. It cares not a whit for our precious scheme of moving step by step from chaos/waste (resource) to our projected image (the paradigm returned or goal achieved). Rather it would have us do as it does—that is, experience whole images melting and forming other images. If we attack, use up, or throw away the image which confronts us before it has metamorphosed into its next stage, we might gain the temporary use of a 'resource,' but in turn lose the game.

'Civilized man' is so programmed that he can only move from chaos, or the unfamiliar, towards what he already knows and values. He is conditioned to turn resources into commodities, environments into deserts. Such a sequence the discovery man holds to be a reversal of the natural order. His attention moves from the familiar towards the unknown, or 'chaos.' For it is only at the moment of turning to the unknown that the wilderness mind, or mind in real time, may appear.

Each discovery man has somehow found that he has no choice but to deal with chaos as prime reality. Within the disturbance of 'things going wrong' may well lie a considerable mystery—a mystery that would communicate itself to him. And probably not just one mystery, but a whole series of them—much like the sequence of colored handkerchiefs, each knotted to the next, that the magician pulls from his sleeve. Every new disturbance that interrupts the thought and perceptions of wilderness man suggests that the great mystery wants to reveal further of itself to him—and through him to others.

The crucial point is to be aware of chance 'irritations.' Normally, we brush them off. But they can be whisperings to pay close attention. If the mystery were not ready to communicate—so the man of wilderness believes—he would not have been disturbed at that point in time and by that particular set of circumstances. We have all experienced in some of our more vivid dreams how certain details are magnified. Our attention is seized by these anomalies and directed with power and clear intent towards the heart of the matter. In fact, so strange can these experiences seem that the imprint of the eidetic message may endure for years. In a like manner the details of a religious painting—a medieval icon, a canvas by El Greco—may also be 'distorted' to entrance us with an aura of holiness, or that which is beyond time as movement.[1]

Disorder-chaos can serve a purpose similar to the heuristic dream. It renders visible the scope and nature of the injuries inflicted on the natural mind by the formatting culture. The traumas by which cityculture marks its initiates are not so visible as were the tattoos and cicatrice designs used by earlier cultures. Outwardly, there may be a change of haircut and dress. More important, however, are the fresh aversions and time-changes scarring the newly accredited city mind. Such inner strictures might pass unnoticed except for moments of challenge. For it is by our inadequate and inappropriate responses to them that the contours of our resistance to perception are delineated. Whatever we need to uncover in order to experience ourselves as a whole human being is again and again thrust upon us. In such moments it would seem that creation is very much relating with us. We know that in our physical development minimum levels of maturity are necessary for the organism to function at all. Without fully formed heart valves, the essential glands, the proper blood chemistry, the infant man perishes early. All functions are important, but some are vital. Likewise with man's psyche. Here, creation expects us to relate to the not-known. And it continues to feed that to us, bit by bit, image by image.

In the wilderness experience, the urban identity—which we have structured for ourselves out of a series of finely tuned expectations—collapses. The familiar is lost. We are faced with situations and images for which we have no correlative words and scenarios. Under this shock our elaborate time-images begin to dissolve and seem 'unreal.' Clinging to them, we will feel ourselves also melting. If on the other hand we should allow them to slip from us with as little regret as possible, we might observe a remarkable phenomenon. As they start to blur and change, new images may emerge out of the old which we have never seen or thought of before. What had been 'real' drops away like the tail of the tadpole. Then, the mind is free to begin its 'blind' exploring of that with which it collided. That we have not the words or the conceptualizations to measure and classify the nature of this not-known matters not at all. In fact, such an element of mysteriousness and 'danger' is essential and can be thrilling. In this organic, ongoing process even the image of what disclosed itself only yesterday may dissolve and today be recast into a still more striking and illuminating image.

A Hindu tale recounts the advice of the sage to the woodcutter: "Go forward, my son!" The woodcutter found sandlewood and started home with a load. But he remembered the admonition to go forward, to go deeper. He toiled into the mountain fastness until he came upon a silver deposit. But with even this wonderful discovery, he did not end his searches. Past the silver he went and came upon a ledge of gold. Again, the words rang in his mind, "Go forward!" finally, far into the mountains and past all tracks of men, the former woodcutter came upon *diamonds*—which, of course, translate to the shaman's congealed light.[2]

Fashioned to explore the strange, man cheats himself and creation if he accepts values established by others in some far off time. Creation is rich indeed. Settling for the familiar is the fate of the worm. The mind of uncertainty, in its mode of ongoing question and search, is more powerful than the mind pleased with the elegance of its conceptualizations.

IN CITYMAN, RECOGNITION, which is non-perception, has been elevated to an ideal. We teach recollection, but not the watch, wait, and poking about that are necessary for detection of the unfamiliar. We are adept at the use of flash cards and the memorizing of dates, historical figures, and foreign words. But, to perceive is something quite different than mere recognition. It is, in fact, just the opposite. To recognize kills perception. And to perceive truly, recognition must perish. Perception is, of course, not presently encouraged, nor has it been for a long while.

This condition of chronic non-perception which is endemic to cityman is quite similar to amnesia. But such loss of mind power is not due to some chance physical injury. Non-perception is an adaptive shield. Behind it we would protect ourselves from that which cityculture chose to withhold from our young experience. This 'playing stupid'—which stems from our acquired aversion to chaos—is part of the psychic mechanism by which cityman is taught to define himself and to limit his extension. He may euphemize this aversion as his taste, as his choice, as his fine discrimination, as his professional expertise, as his ambition, or some other term of recognition. But none of this is perception, or a true discovery of the wordless. Any real confrontation with the not-known throws our mind into the timeless: yesterday and tomorrow vanish. And our inner monologue—that singsong of traditions and hopes we have been encouraged to think is our mind—comes to a dead stop. Such a suspension of conventional time (time as the endless belt moving) is experienced as a momentary paralysis and a fright. Seen another way, confrontation with the unrecognizable is the pressure of our individual matrix-intelligence to express itself. Perception would displace recognition.

By this phrase, *matrix-intelligence*, we mean that impulse to full maturity which the oak tree 'knows' in the acorn. All sorts of influences will affect the seedling and some may prevent it from attaining the stature of patriarch of the forest. But implicit in each acorn is such a metamorphosis. Speak, if you will, of DNA and genetic codings. The conclusions are the same. Each start is already a complete manifestation, injuries along the way notwithstanding. This matrix of intelligence is the integrity of the organism. What is true of our development in the physical and familiar also holds true in the paraphysical and unfamiliar. Rudolph Otto wrote in his seminal work, *The Idea of the Holy*:

> Above and beyond our rational being lies hidden the ultimate and highest part of our nature, which can find no satisfaction in the mere allaying of the needs of our sensuous, psychical, or intellectual impulses and cravings. The mystics called it the basis or ground of the soul.[3]

This "ground," or matrix, will not permit us to play stupid. We may want to retire to a small comfortable circle of familiarity in which the impingements of life are limited and events severely controlled. But that is not in the mind of creation. Trying to achieve what we already know only cubicalizes us. The invocation of a goal blocks the activity of discovery-perception. That the goal in mind happens not to be in sight alters the situation not at all. Achieving is mere recognition carried to another dimension. To be sure, an activity of the mind does ensue. But it is not a 'forward' movement as is often claimed. To recognize is to retreat, to deny, to avoid seeing/listening/sensing acutely. Reality in creation's sense of that word is not recognition, but a simple turning towards the not-known. The effort to concretize one's field of familiarity is a departure from what is genuine. We live by encounter with the not-known. We die smothered in the winding sheets of the recognizable.

The discovery man willingly risks the adventure into the unknown, uncertain of and unconcerned over the outcome—even if it ends in his so-called failure or death. He does not project himself into the unknown, as he might plan a trip to Paris. He cannot achieve the unknown by sitting down and throwing his mind into nothingness. It is not a matter of contemplating one's navel. The unmanifest is not nothing. It is, indeed, all that which lies beyond what we recognize. To find oneself in the rich jungle of the undiscovered and unnamed is electrifying. Nothing in cityculture amusements or spectacles can equal such a vivid, quantum acceleration in the mind-brain. (In our essay entitled *Life Of The Mind* we shall examine many splendid examples of this exploratory mind at work.)

No goals, no charts, no traditions guide the wilderness wanderer. He must, if need be, grope as in the dark, stumble and crawl like a baby, even fight for his breath like a newborn, and bear whatever hurts may go with his wilderness experience. All these discomforts are incidental. For him there is no living beyond participation with creation. Life becomes then the emergence of what is beyond the cultural known—what few men, perhaps none, may have yet perceived. To such an individual creation gives its own initiation by vouchsafing a small part of itself to him.

THE DISCOVERY CYCLE

THE WILDERNESS MAN and his discovery mind have come in every guise and intensity. The range of the expression has been from the sublime and total manifestation to the fragmentary. Even the disturbing vision-like dream—intruding into an otherwise seamless kingship or traditional kinship identity—must be noted as a break-in of the wilderness experience. This reset of the mind into an unstructured time area between tradition and hope, between expectation and anticipation, has, on many occasions, been described as a religious experience, or at least as a 'philosophical speculation.' In the man-creation relationship we have had many prime discovery minds, among them a Buddha, a Jesus, a Shankara, a Ramakrishna.

At other times and places this special wilderness-facing has arisen in the relationships between man and his environment. Then we have had the 'bringing' of fire by a Prometheus (or a Lucifer); the investigation of herbs and barks and roots for medicine by nameless millions; the invention of the keel by some unremembered Norse geniuses which enabled ships to sail against the wind for the first time ever. Metallurgy, the orbiting of the planets, mathematics, and a thousand other investigations of the not-known were each at some one point in time the result of this nontraditional activity. For not one of these insights by which we live today could have been brought about originally through the achiever's 'directed thinking' nor through the recall of former traditions. Each represented a stoppage of the familiar and then a breakthrough of fresh insight: some human consciousness troubled with a 'difficulty' was seized with a thrilling wordless experience. And a perception of a new reality poured into his mind as though he were being anointed.

Such 'increases' have been portrayed by cityculture as resulting from evolution and achievements. But the phenomena which the insight rearranges in the discoverer's mind had already been present and active for perhaps hundreds, thousands, even millions of years. Nor does man himself seem that changed in his essential powers over some of these long time frames. In the far past he simply asked one kind of question; yesterday, another; and today, a few of our minds linger over still other curiosities. And, we may note, many of the 'original' insights which proved basic to man's survival were made by the early hunters—without benefit of libraries and research stipends.

Certain fundamental problems which worried earlier people do not seem to bother us anymore. And that is not because we have solved all the questions they faced. In our wisdom we try to forget or make of each such enigma a non-problem, or a nursery story.

Take the Sphinx and her riddle. Travellers to ancient Thebes had to answer the goddess correctly or risk death. The problem, we are told, was, "What creature goes in the morning on four, at noon on two, and in the evening on three?" Why the question appeared so formidable to the ancients seems puzzling to us now. The approved answer, we were told , was "man." That reply, however, was insufficient and would have gotten you in serious trouble. The Oedipal legend notwithstanding, the Sphinx had posed a much larger question: "Explain, if you can, the nature of time."

The riddle of the Sphinx is still unanswered and just as demanding of a response as it ever was. But if we are not even aware of these matters, can we say that we have progressed very far from those people who were concerned? In this way, the discovery experience cannot be perceived as the late climax of evolution. Rather, it was well known to early man and all but forgotten by contemporary man. On the whole we avoid that state of mind which inquires and wonders, for it also entails engaging with the unfamiliar and the uncomfortable. The discovery mind is our unalienated mind. Its function is an inevitability which can be put off no longer.

The most spectacular and memorable displays of this discovery mode have probably been the shamans, prophets, and avatars. They were such expressions of the timeless mode of the mind that their powers of intuition were often perceived as magical. They seemed endowed with the ability to call on creation directly. Consider the legends depicting the wonders of Elijah's mastery. His Lord was one of the new style sky-gods—those kingly Bronze Age controllers of wind, thunder, lightning, and the uncertainties of tomorrow. The peasant populations, we are told, still hankered after their older earth spirits and fructification deities who, of course, governed the great calendaric events. Elijah challenged the priests of Baal and Asherah to a shamanistic contest.

> You shall invoke your god by name and I will invoke the LORD by name; and the god who answers by fire, he is God.[1]

The priests of the chthonic deities danced and chanted for hours without igniting the sacrificial pyres. Elijah's taunts sound almost modern:

> At midday Elijah mocked them: 'Call louder, for he is a god; it may be he is deep in thought, or engaged, or on a journey; or he may have gone to sleep and must be woken up.'[2]

Then, we are told, Elijah called on his sky-god, master of the new sources of power. (Note the importance given to the timing of this magic.)

> At the hour of the regular sacrifice the prophet Elijah came forward and said, 'LORD God of Abraham, of Isaac, and of Israel, let it be known today that thou are God in Israel and that I am thy servant and have done all these things at thy command. Answer me, O LORD, answer me and let this people know that thou, LORD, art God and that it is thou that hast caused them to be backsliders.' Then the fire of the LORD fell. It consumed the whole-offering, the wood, the stones, and the earth, and licked up the water in the trench. When all the people saw it, they fell prostrate and cried, 'The LORD is God, the LORD is God.'[3]

Clearly, these unusual, nontraditional individuals astonished and frightened those whose minds were locked into what was then the familiar. And how more dramatically could the time-shifting powers of the wonderworkers be described and recalled by their peasant followers than by apotheosis? These minds lived and moved in a time different than used by most. Legends have them fetching lightning bolts from the sky. But more importantly, they were capable of summoning the powers of real time.

Such unusual characters did live, however, and their minds were capable where ours are not. Some have walked on coals of fire. Some have healed men of their diseases or infirmities. Others have read men's thoughts, intuitively perceived events at a distance, or gathered hungering multitudes to them like a mother calling her children. For us such minds are nonstandard and subject to being diagnosed as pathological aberrations. We find it nearly impossible to accept the possibility of minds other than our own tripartite, self-opposed achiever minds. And a mind able to function in the unfamiliar, or the narrow time slot between yesterday and tomorrow, staggers us every bit as much as it did the peasants at Elijah's demonstration. We cannot see how to enter into such a mode directly ourselves. Nevertheless, the power of the mind in its wilderness-turning mode is inherent. In spite of its being covered over with aversions early in our childhood, it still remains an inalienable faculty. Our dominant culture—and its tradition-achiever identity system—may prevent us for most of our lives from detecting, much less experimenting with, the discovery mind. But exploration of the unfamiliar is our nature. Whenever and wherever there is a breach in our conventional identity systems based on tangibility, the thrill of the 'dark' and unperceived can rise up out of nowhere to set the hairs on our neck tingling. Life seems charged with infinities to consider. At such moments the individual senses that he is an ongoing, open-ended process. And that process has nothing to do with rearranging the familiar, but would forever be digging further into the not-known.

Such a hiatus, or little death of the usual, can come about in many ways. We lose some essential thing or piece of information. Despite all our searching we cannot find it. As a result the whole carefully balanced structure of our identity-life also seems to vanish. We are at that moment—for a second, for a few hours, or for a few days—as though in a wilderness where there are no trails and no signboards. We have no idea of how to proceed. This very lack, or blankness, or failure of the familiar, forces our mind into a different mode. Our ability to achieve or to move towards a goal is frustrated. And our ability to recall tradition (where the lost item ought to be) is thwarted. In this impasse the ordinary will not work. We may grow angry, become heated, throw things, and yell. In Biblical times men looking for their lost donkeys might ask a seer to use his intuitive powers to find the missing animals. Or a Sioux family might ask a *yuwipi* man to locate a lost child.[4] For us the lost usually stays lost until we have given up on it. Then, sometime after the 'giving up,' it may suddenly appear as though it found us. The name of the forgotten friend drops into our mind. The missing bottle of aspirin we searched for everywhere was all the while beside our cup of coffee.

The momentary discontinuities we experience daily are miniature, but nevertheless very real experiences at the edge of the unfamiliar. It is as though the sleeves and the hems of our coats were ever being plucked at by the slipstreams of that powerful vortex into which the discovery-man has fallen altogether. The wilderness mind is not at some distance, but immediate. The difference between us and the discovery man is only a matter of intensity and of time. Faced with a ripple or full disturbance in the continuity of his existence, the discovery-man does not try to counter it with a retreat into a more familiar time; he does not reach for an antidote of fundamentalist wisdom or achiever-action. Instead he hangs in the 'suspension of time' and applies his most wakeful attention to the nature of the disturbance. He does not try to defend the familiar image which seems about to dissolve, but listens and watches intently for what is taking place.

That which triggers the wilderness experience—of whatever degree—is, therefore, this very common, very human frailty which we all encounter every day. Our plans come unglued. The fabric of familiarity tears. Our frame of expectations begins, little by little, or all at once, to distort. The opener can be frustration, disruption, incongruity, misunderstanding, fascination, or shock. But quite often the most significant and choice experiences with the unknown may be presaged by no more than a delicate wavering of the familiar. It can be a disturbance or discontent so faint that we scarcely notice and hurry on by. A quite common introduction to wilderness is the loss of something which we must have to continue an even, controlled existence. That something could be our way in a new city. Or, the loss could be a special tool upon which a whole chain of important events depends. Or, it could be a sense of our own worth. Or, it might even be the accountability and 'reasonableness' (the good intention) of creation itself.

In John Muir's account of his early rambles in the California Sierra we come upon a full-blown wilderness experience in a true wilderness setting. He has gone off by himself to explore a mountainous area and hesitates to climb further up a rugged and inviting peak. "But we little know until tried how much of the uncontrollable there is in us, urging across glaciers and torrents, and up dangerous heights, let the judgment forbid as it may."[5] He pushes on, scrambling up cleavages and over rough blocks, enthralled at the widening scene. He reaches a perilous traverse. He cannot go down and further progress seems impossible:

> At length, after attaining an elevation of about 12,800 feet, I found myself at the foot of a sheer drop in the bed of the avalanche channel I was tracing.... [T]he walls were smoother than the obstructing rock, and repeated efforts only showed that I must either go right ahead or turn back. The tried dangers beneath seemed even greater than that of the cliff in front; therefore, after scanning its face again and again, I began to scale it, picking my holds with intense caution. After gaining a point about halfway to the top, I was suddenly brought to a dead stop, with arms outspread, clinging close to the face of the rock,

> unable to move hand or foot either up or down. My doom appeared fixed. I *must* fall. There would be a moment of bewilderment, and then a lifeless rumble down the one general precipice to the glacier below.
>
> When this final danger flashed upon me, I became nerve-shaken for the first time since setting foot on the mountains, and my mind seemed to fill with a stifling smoke. But this terrible eclipse lasted only a moment, when life blazed forth again with preternatural clearness. I seemed suddenly to become possessed of a new sense. The other self, bygone experiences, Instinct, or Guardian Angel—call it what you will,—came forward and assumed control. Then my trembling muscles became firm again, every rift and flaw in the rock was seen as through a microscope, and my limbs moved with a positiveness and precision with which I seemed to have nothing at all to do. Had I been borne aloft upon wings, my deliverance could not have been more complete.[6]

Although Muir's adventure may seem remote in time and place—we now hike on graded trails where once he risked his life—this finding ourselves spread-eagled on a 'precipice' is not extraordinary, but basic to the human phenomenon of experiencing. When we are really locked into the wilderness initiation, we will not be able to find our way back to normalcy with any amount of reason or money or even with new ideas we get from others. The situation becomes an impossible double bind for us. The ordinary and familiar no longer work at all. We feel forced to 'make a move'—*any move*—but every choice appears fraught with as much trouble as we face where we are. There is no 'better tomorrow' to which to escape. Our conventional mind becomes totally blocked, useless, at a dead end. In the 'onset of the mode' we may even seem headed for some kind of madness. Jung describes such an encounter with the uncomprehended and quite candidly confesses this impasse with the unfamiliar to be beyond cultural remedies:

> In actual reality we do not have at our command any power of cool reflection, nor does any science or philosophy help us, and the traditional teachings of religion do so

> only to a limited degree.... A turbulent life-situation has arisen that refuses to fit any of the traditional meanings assigned to it. It is a moment of collapse. We sink into a final depth—Apuleius calls it "a kind of voluntary death." It is a surrender of our own powers, not artificially willed but forced upon us by nature; not a voluntary submission and humiliation decked in moral garb but an utter and unmistakable defeat crowned with the panic fear of demoralization. Only when all props and crutches are broken, and no cover from the rear offers even the slightest hope of security, does it become possible for us to experience an archetype that up till then had lain hidden behind the meaningful nonsense played out by the anima. This is the *archetype of meaning*.[7]

Jung describes the crash of the contrived personality against the uncomprehended with all the familiarity of personal experience. The "archetype of meaning" is simply his oblique way of saying "insight."

This breakdown of the usual—never really expected or welcomed—will infuriate most of us. In extreme situations the loss of 'order' will appear dangerous and a collapse of 'civilization.' Yet a *cycle* of four stages can be distinguished in these anomalous, wilderness experiences. In our disturbance we are usually not aware of these phases, but the discovery man notices them minutely as essential to his initiation.

First: the mind experiences a frustration of the usual. Initial efforts at restoration fail.

Second: while others still relate comfortably to the familiar, the "unlucky" individual must deal more and more with the uncertainty which dogs him. He becomes obsessed with what he cannot define. He searches everywhere for "answers" and "solutions." Others perceive him as odd. No one can advise him. He finds himself isolated; the problem for now is his alone.

Third: the mind comes to a dead halt such as it rarely experiences. The internal monologue switches off. The silence is terrifying. All useful ideas and hopes are exhausted. The individual finds himself immobilized in the painful shock of his blankness. The world that he now discovers is one for which he has no words or 'explanations.' At every turn what had been familiar is now altogether strange and hostile. Where he had pursued his torment vigorously, he may grow listless. He feels himself sleepwalking; his mind loses focus. If he responds at all, it will be to whatever randomly stirs his curiosity.

Fourth: minutes, days, weeks, or even years later, an unexpected and wholly new perception of his difficult situation rushes in upon him. The mind's sudden grasp of a greatly expanded 'map' is experienced as a kind of rebirth. With the 'unlocking' comes a sudden euphoria and sense of worth unlike anything a man-made culture can offer.

Finally: what seems a 'fifth' stage in this discovery cycle emerges as the individual attempts to comprehend more fully the nature of his 'insight' and to share it with his community.[8] His mind stumbles for words to encompass the wonder. Then he realizes he has not yet found his way through the unknown: unexpected questions and anomalies have shown up.

The 'fifth' stage is the cycle starting over again: frustration of the usual, psychic bafflement, close attention-investigation and then an inrush of unexpected new images. With each cycle he may feel he has finally wrapped himself around the "problem." But he will not know until he tries to explain or write about what he has found.

Job waxed eloquent in defending his preconceptions of holiness through attention to ceremonies. He demanded to confront God. He cried out: "Let me but call a witness in my defence!/ Let the Almighty state his case against me!"[9] Later God deigned to reply: "Who is this whose ignorant words/ cloud my design in darkness?/ Brace yourself and stand up like a man;/ I will ask questions, and you shall answer."[10]

Here we have a formula for the wilderness experience. We cloud creation with preconceptions we call words. Man would use these words to bind creation into the familiar, to cut up time and make it manageable. And we expect creation to answer us in terms of our preconceptions. But it is creation that is the question-asker, the problem-setter, the puzzle-maker, the Sphinx. Creation is the game-player and riddles us. As we respond, if we respond, we live and 'grow.' Through a kind of four-stage psychic metamorphosis, our grasp of the world in which we float undergoes a continual reorganization. We incorporate ever increasing perceptions of the radically new. All that creation can share with us is a series of wonders. These, if allowed to take hold of our minds, can break our preconceptions and, through an image-into-image process, create us as whole men, braced and standing up.

IN SEARCH OF THE UNCOMMON MAN

HISTORY AS SERVED TO US in newspapers, magazines, school books, and patriotic addresses has a perverse way of leaving out a class of men and women most uncommon. They had no titles, they had no power. You won't find them mentioned prominently in Creasy's *Fifteen Decisive Battles.* Not one of them made an all-night gallop with troops to stem the tide of the 'barbarians.' They converted no tribes to slaves, they built no temples. Nor will you find one astride a raring horse, immortalized in bronze at the center of a town square. Even so, history would not be as we know it were it not for these special, little noticed individuals. Although they made no efforts to coerce events, history has nonetheless been forced to bend around them. For these uncommon individuals were each the source of new definitions, new ideas, new discoveries. And in ancient times such innovators were long remembered as the gods and goddesses who brought special boons to mankind.[1]

The appearance of the uncommon man and the wilderness mind have taken many forms. Among the dawn tribes—the hunting and gathering people of prehistory—the discovery-mode was to be found in the uncommon person of the shaman. In the transitional agricultural hierocracies and the feudal kingships the discovery man might be the prophet at the gate, the philosopher, the alchemist, the innovative nun, the disputatious monk. (The priest, we note, was rarely an uncommon man. As a creature of the dominant culture, he was usually a role-player and an apologist for the traditional.)

The persistence of the wilderness-turning faculty of the mind is that of a wind-battered juniper anchored in whatever cracks are available. In the earlier citycultures of the Mediterranean and Europe, the wilderness mind cropped up again and again to affect the parameters by which man thought of himself. We might include the legendary Empedocles and Pythagoras, and investigators of the mind and values such as Socrates and Diogenes. Much later, the son of a silk merchant retreated to a ruined chapel and searched for a way to restore an experience of what is not known—often thought of as the holy—to the individual. Saint Francis's movement of "Little Penitents" was "an awakened enthusiasm for 'soul-freedom,' or a protest against the tyranny of church and state, an outbreak of spiritual and communal emancipation. It was a holiday of the heart. The people, like children, were let out of school and rushed forth into the fields to play, sing and dance...."[2]

It was in this climate of expanded, unlocked sensibility that Giotti brought in his new way of looking and recording, and it was with his eyes that the western world afterwards saw itself. Others followed, perceiving the world not through tradition or culture, but as freshly discovered in their own manner. Brunelleschi changed architecture and showed how the illusion of perspective could be realized. Donatello sculpted friezes of spirited men and scenes like none before. Jan Van Eyck and Holbein dared us to examine our world exactly for its content of truth. And in the realm of thought

Wycliff, Huss, and Luther questioned the established mind-control powers. With a force that no prince or magnate could have achieved, their challenges shook the known world for centuries thereafter. Later, other men, without leave or authority, but with their own wit and insights, redrew the cosmos—Copernicus, Kepler, Newton, and Einstein. Finally came poet-artists such as Blake, Wordsworth, Coleridge—men of an inner vision such as Jakob Boehme, working outside the envelope of traditional thinking—who would return us to validation by way of the wilderness quest.

Such psychology as we have is incomplete. We find no body of work which investigates and describes the exceptionally fertile mind of the uncommon man. We have little practical speculation on why our everyday minds do not experience a like access to insight. What we have in our city psychology, then, is something less than a science. This discipline is concerned only with the standard achiever mind and how to gear down the natural mind (or, as is said in educational circles, "dumb down") to approximate the tripartite, self-opposed, choosing mind. If we were to have a genuine psychology, it would have to include those minds which have worked beyond the edge of the familiar. It would have to describe and account for their dynamics, especially those attitudes and faculties which seem entirely missing in our minds.

The uncommon men seem always to have stood apart—not as aliens, but each, as it were, as his own solar system. Like ourselves they are subjects of this country or that. But, possessed of an aloofness to the hysteria of the predominant culture, they are feared by the current kingship, temporal or secular. We, on the other hand, are products of whatever culture in which we happen to find ourselves. The minds we use are borrowed. Indeed, it cannot be said that we truly have a mind of our own until we have had our own validating, wilderness experience of time. So long as we feel some comfort in a sphere of familiarity defined entirely by others, we shall be serfs of the time embedded in the mind we have on loan. Our thoughts and actions can be nothing other than the working out of formulae and parameters beyond our reach and influence.

Our present minds, being functions of the culture, are vulnerable on demand. At any moment, day or night, waves of hysteria emanating from our kingship leaders can trigger response mechanisms over which we have little or no control. Without our own experience of time we are helpless. As a consequence we can be rolled over, turned this way and that, and sent marching against strangers even to our death on a battlefield under the 'right' flag. As clones of the culture, we are not using the natural mind in us which would give us insight and discrimination. The uncommon man, on the other hand, has been validated in time by wilderness itself. He 'thinks' without the familiar hopes and expectations with which cityculture tangles and traps us.

ONLY RECENTLY HAVE WE BEGUN to recover evidences of man's mind *prior* to both the achiever and the earlier agricultural eras. Ancient libraries long lost in the deserts of Egypt and the Ancient Near East have been recovered and the rare texts translated. Caves of the reindeer men of Europe circa 18,000-20,000 B.C. have been located and their special libraries, walls covered with murals, have been meticulously recorded. And we have preserved a few absolutely priceless conversations with our own genuine reindeer people just before the extinction of the great hunter-gatherer tradition from the world. The several Black Elk volumes and *Teton Sioux Music* are among these latter treasures. Prehistoric France, Roman Palestine, 19th-Century American Far West, all once shared a phenomenon of great vitality. Widely separated in time and space, they witnessed astonishing exhibitions of the wilderness mind. In all this diverse, recovered material we happened upon ideas and psychic states which we earlier found in the Bible. Some of these strange stories had seemed more myth than recounting. Because we have recently come into possession of second and third sources on the Biblical era and its ethos, many scriptural passages which have long puzzled us are less difficult.

Once thought divine, the uncommon man can now be perceived as the survival of a former way of defining time. He is the late expression of a special tradition in all likelihood as ancient as man himself. In accounts of the uncommon men (and women) we notice that they invariably experienced, usually as they neared adulthood, a period of disorientation. Well before the appearance of their shamanistic or wilderness self, they sometimes fell strangely sick and listless. Among the hunters, the young individual so affected might wander into the wilds by himself, living as he could on what he found and on what his family brought from the village or camp to him. In all appearances he would seem to have lost his mind. (Today, we would doubtless subject him to insulin, or electroshock treatments, to "restore his hold on reality.") But early peoples were wiser than we in these matters. They understood that the individual was undergoing a series of psychic changes—an important subjective metamorphosis. He was learning to go beyond the bounds of the culturally known—the screen of words—and relate nakedly with the subtle and the unmanifest. Recovery was not certain. But those who did return were often changed most remarkably; they had wandered off as children, but they returned as adults with powers that augmented the community.

Those young people, on the other hand, who were never spontaneously infected by such 'sickness' would become adults of the community by another, but similar route. So important was the actual wilderness experience—this ability to function outside the familiar or home territory—that some tribal communities would attempt to induce a confrontation between the young individual and the not-known. He might be sent off into wilderness alone for a period of days. He might be left in a cave to experience there the whispers of the unknown. And some tribes, we know, made use of the hallucinogenic powers of Jimsonweed, peyote, or other naturally occurring alkaloids to generate the semblance of a boundary-crossing experience.[3]

The one who returned from the wilderness, after having wandered in a state of disorientation and unfamiliarity, became an adult of a special degree. His initiation into the mystery of time was his own and from creation directly. His rapport with the environment appeared phenomenal. Filaments of perception seemed to run out from him in all directions. What he could not see or hear, he might sense intuitively.

The early hunter understood that it was all right to be lost. Being lost without panicking was essential to his survival, and in turn to the survival of the community. Man evolved in an incredible world of many mountains, rivers, oceans, plants, birds, animals, fish, and other people. Whole lifetimes were spent in experiencing and noting the fine degrees by which creation varied. Hear how a Teton Sioux shaman, named Shooter, "a thoughtful man and well versed in the old customs," expressed the hunter's concern for the concrete, the individual, and the genuine:

> The forest is the home of many birds and other animals, and the water is the home of fish and reptiles. All birds, even those of the same species, are not alike, and it is the same with animals and with human beings. The reason Wakan'tanka does not make two birds, or animals, or human beings exactly alike is because each is placed here by Wakan'tanka to be an independent individuality and to rely on itself.... From my boyhood I have observed leaves, trees, and grass, and I have never found two alike. They may have a general likeness, but on examination I have found that they differ slightly.... A man ought to desire that which is genuine instead of that which is artificial.[4]

The genuine for the hunter was the minute difference that denoted change and individuality. His eye was attracted to the unfamiliar, to what escaped definition. To accept his surroundings as wholly familiar, failing to detect the sham look-alike, was to be rendered vulnerable.

Early men depended for their continuance as a hunting-gathering kind upon an intimate relationship with whatever environment they were in at the moment. Being uncertain, being unfamiliar with what confronted one, was a normal, even everyday experience. Long before the first agricultural villages and the hierocratic cities, mankind had spent many hundreds of thousands of years moving freely about in the strange and wonderful. In their hunting men were always encountering the unexpected—becoming lost, as it were, exploring, and finding themselves again. When they did, the old physical boundaries and space realities had been replaced with a new and enlarged arrangement. They returned not the same men as had set out. Analogous to uncertainty of terrain was the experience of becoming psychically lost, of one's sense of everyday time dissolving. On recovery, one returned a different man, a shaman who could travel in a timeless, measureless, wordless space.

Among the first anthropologists to defend early man from the denigration heaped on him by the Freudians and the popular press was Paul Radin. In *Primitive Man as Philosopher* he described the particularity with which early man considered his environment:

> An object is *not thought of as the sum of all the sense data connected with it.* A mountain is not thought of as a unified whole. It is neither static nor is it a series of inherently connected impressions. It is a continually changing entity from which one is repeatedly subtracting and to which one is repeatedly adding.... To talk of a tree being the same when it is constantly undergoing transformations is based on an assumption which the man of action simply does not make.... He conceives the possibility of imagining it having an entirely different appearance on the following day.[5]

Creation is a complex field of many energy movements. It behooves man to be aware of subtleties of change at every level. Whatever has been labelled familiar and made static through a word may be false in this reality of shifts and variations. Again Radin:

> We are accustomed to derive all our proof for the existence of a thing from the evidence of our senses. The cultured man of Western Europe is, in the main, as we all know, visual-minded. That some inward feeling or stirring, some sudden and vague sensation or intuition, might be taken as real proof for the existence of an event would not occur to him.... No one of us, as far as I am aware, would seriously contend that an inward experience—the presence of an inward thrill—could establish the reality of the whole cultural background. Yet this is precisely what does happen in primitive culture, for the man of action. Why, so he would contend, should something affect him in this way, if it were not true.... This is to him as much of a real proof as anything happening outside of him.[6]

And finally:

> From the man of action's viewpoint, a fact has no symbolic or static value. He predicates no unity beyond that of the certainty of continuous change and transformation. For him a double distortion is involved in investing the transitory and ceaselessly changing object with a symbolic, idealistic, or static significance; first, because we then remove it farther from reality; and second, because in thus separating the perceiving self from the object we really render both of them meaningless.[7]

Every young person on approaching the threshold of adult life is affected by a panic of one degree or another. He may be quite conscious of his unease and the source of it. More likely, he will simply feel sick. The crisis that confronts him is the problem of living in a field of preconceptions. Are they real? How do they work? Will they work for him even though he had nothing to do with setting them up?

The young hunter feeling this deep unease with the familiar and turning from camp to wilderness for his own personal initiation as an adult in time was not resigning from the community as an incompetent, a neurotic, a weakling, a failure. Quite the contrary: by withdrawing for a season from the everyday, he was walking the difficult, unmarked path. He was taking on the work of the primal adult, inviting man's most basic and challenging experience—the direct, *wordless* relationship of his point of energy to the whole. If he is favored, he will have recapitulated in his psyche the final stage in the metamorphosis of man. Joseph Campbell in his excellent *Primitive Mythology* has given us these valuable observations:

> The shamanistic crisis, when properly fostered, yields an adult not only of superior intelligence and refinement, but also of greater physical stamina and vitality of spirit than is normal to the members of his group. The crisis, consequently, has the value of a superior threshold initiation: superior, in the first place, because spontaneous, not tribally enforced, and in the second place, because the shift of reference of the psychologically potent symbols has been not from the family to the tribe [as most often in the agricultural communities] but from the family to the universe. The energies of the psyche summoned into play by such an immediately recognized magnification of the field of life are of greater force than those released and directed by the group-oriented, group-contrived, visionary masquerades of the puberty rites and men's dancing ground. They give a steadier base and larger format to the character of the individual concerned, and have tended, also, to endow the phenomenology of shamanism itself with a quality of general human validity, which the local rites—of whatever community—simply do not share. And finally, since the group rites of the hunting societies are, *au fond*, precipitations into the public field of images first experienced in shamanistic vision, rendering myths best known to shamans and best interpreted by shamans, the painful crisis of the deeply forced vocational call carries the young adept to the root not only of his cultural structure, but also of the psychological structures of every member of his tribe.[8]

Descriptions of the shaman's experience from diverse areas of the world depict men and women on the threshold of the unknown. To convey the dramatic change in his nature, the new shaman might resort to hyperbole, as we have noted, and tell how his bones were cleaned, counted, and reassembled; of having his body stuffed with rock crystals. How much more vividly than this could one convey the sense of being stripped and scraped clean of the familiar and preconceived? The jewels are light caught and "congealed."[9] One's inner being is no longer the darkness and anxiety of alienated time, but has become a place of real-time brilliance and vision.

Such a special 'illness,' followed by the recovery that changed and strengthened the individual, was once an integral part of man's understanding of his psyche. Evidences of this initiatory experience of the hero-shaman can be found not only in these gleanings from the last of the hunter people, but still embedded in our own folklore. The Grail Quest, the Forty Days in the Wilderness, the Voyage of Ulysses, the Trials of Hercules, the 'Perils' undergone by Celtic heroes were all ventures of individuals outbound from the familiar. They sought direct *mano a mano* confrontations with the unknown over the meaning of time. In every age we can find marvelous examples of the uncommon man who makes a perilous passage—between clashing rocks, over a razor-like bridge.[10] He must enter the realm of the unknown through the hairline crevice between yesterday and tomorrow. Not until he has experienced the not-known will a man became fully a man.

Among the early hunters, the ones who returned from wilderness lived in the community much as others did. But ever afterwards such a one might experience 'lapses.' With little warning the state of mind which he had discovered in wilderness, and which was peculiarly his own, might assert itself. A queer feeling would steal upon him and his mind would fall into a state of timelessness. Out of that he might describe some of the hints he felt. In the earliest cultures this uncommon man, who could live now and then in creation's timeless time, was accepted and relied upon by his kinsmen. As the hunting communities, however, were displaced by the planting and cultivating communities, this uncommon man became feared. In the later, historical periods we might find him living at the borders, in caves, with the nomads, in the forests, wherever wilderness and the timeless still held their own.

Although the hunters were routinely exterminated and the few survivors driven into evermore remote jungles and mountains, the validation of man's mind through the wilderness experience did not die out. It has continued in one guise or another in all later cultures. The metamorphosis of man's psyche in these latter-day examples proceeded as before through four characteristic steps. First came the initiatory sickness and breaking down of what had been familiar. Then followed a wandering in wilderness, or living in the timeless. The individual's 'recovery' came about through his interaction with the strange. Images resolved into images towards a gestalt which he could not have preconceived. On his return, he was possessed of special intuitive powers. As an adept, he would move in community time, but inwardly continue to live in the immediate, the timeless—sensitive to the fine hints to which others were deaf.

THE SOLITARY QUEST

THE UNCOMMON MAN WE HAVE MET as the solitary hero in Greek mythology, the Bible, and Celtic legends. He has existed for us not as a tangible, historical figure, but as the special individual in an indeterminate, mythic past when the suspension of natural laws and bending of time seemed not altogether uncommon. The hero in those remote days might not only be marvelously strong and brave, but endowed with magical powers. When necessary he could call on mysteries with which to overmaster the familiar and threatening. At the outset of each story about an uncommon man—be he seer, avatar, innovator, or discovery man of whatever sort—we invariably come upon scenes of his early estrangement. We discover him wandering outside the conventional.

The classical story of the uncommon man begins with the royal infant being abandoned on a mountainside (set adrift in the great river or out on the sea) and exposed to the elements. Or, he is driven from the court or from the tribal tents by his elder brothers. Or, he is the young man who can become neither merchant-banker nor priest and secludes himself in a hut of twigs in a nearby woods. Whether he is left by his parents, or he voyages in search of a magical fleece, or he flies in quest of the head of an angry, displaced goddess, or he journeys across mountains searching for the whereabouts and secrets of a powerful god, or seeks a Holy Grail or magical cauldron, the uncommon man of myth lives for a while beyond the metaphorical, experiencing whole images. In the mountains or in a forest nearby he exists very simply, eating whatever comes to hand—living in a state of wordless astonishment, or wilderness-chaos, if you will—bereft of a 'normal' identity.[1]

The world for this wanderer in wilderness has turned upside down. Reality and unreality have suddenly exchanged places. The familiar and everyday are illusory. The unnamed—that is, wilderness—now seems the all. Somewhere, just beyond the margins of man's everyday mind, great things are happening. He can sense the power flows which the eagle, the bison, the elk, the snake, the coyote, the bear represent. What is man if he cannot feel his way into an intimacy with this superior reality that looms behind the facade of the everyday? He hungers for that which seems nonexistent to most. This reversal, or breakdown of the familiar, has ever been the prerequisite threshold to the appearance of the intuitive mind which can live and travel in the nameless.

This living as one lost in wilderness, venturing as a new born in the unrecognizable, unnamed, undescribed, has everywhere been characteristic of the discovery-shaman experience. One such remarkable case was the Indian saint, Ramakrishna, who remained in a coma for six months. He was kept alive only by the daily ministrations of an unknown monk who had appeared when needed most desperately. Afterwards, this sadhu wandered on, his name and identity unrecorded.[2] Black Elk, as a child and afterwards, fell into similar 'long illnesses.' It is essential to his lonely novitiate that the wanderer learn to live outside conventional time and familiar identity pursuits. In another era he might have been found sitting with his bowl beside the trail, neither holding on to his life nor pushing it away—waiting for creation to feed him, or to take his life back again as it willed. If one of us had happened upon this ragged, unkempt, even starved individual, we would have thought him either defective, crazed, or moonstruck.

This latter phrase, "moonstruck," is one of those curious and important survivals of an ancient tradition. The moon goddess—'the queen of heaven,' Diana/Isis, the White-Goddess—was likely one of man's earliest deities. The hunters of the prairies knew her as "Beautiful Woman," or "Buffalo Woman." We have even found her as a paleolithic Venus carved in bas-relief on the wall of a Dordogne cave. Here, this *Femme à la Corne* holds up what appears to be a horn, but could also suggest a 'rib' or a new moon.[3] Further to the east such a female deity is to this day worshiped as Kali, the Hindu mother/destroyer goddess of time, who gives birth to mankind, but who, like Saturn, also devours her offspring. To be moonstruck, then, has from earliest times suggested ecstatic rapport with the demiurge of our corner of creation.

Caves, caverns, recesses in old trees—all were used to minimize, even turn off, the material senses, tuned as they are to the everyday, and to give the searcher for the unnamed a chance to hear the unspoken, to see the unseeable. Prospective shamans in Alaska walked in pitch darkness towards a cliff. The problem was to sense by inner feeling, or intuition, the entrance to one of several caves. If he struck the cliff, it was demonstrated he was without the power: the cave had not 'opened' for him. But once inside the cave he would further his adventure in the intuitive life, wait for his illumination, and then escape before the cave closed forever. The great painted caverns in southern France and northern Spain may well have been similar places where the young mammoth hunters sought the shaman's inner verification. As his small oil lamp flickered lower, the footsteps of his mentor fading and the images of the great beast-powers dancing in the shadows, the young hunter waited expectantly. One can almost hear him chanting for a while, as did the Teton Sioux shaman: *"Hee-ay-hay-ee-ee!* I am sending a voice! Hear me! *Wakan-Tanka,* Grandfather, behold us! O *Wakan-Tanka*, Father, behold us!" But as the infinite, solitary, numinous darkness descended, every nonphysical sense of the hunter would heighten to a preternatural awareness. For his initiatory visions of light and power, he challenged, he beseeched creation to explain itself.

Of the ancient's quest for the nameless, everchanging-unchanging, Eliade has observed, "Everywhere, we find the will to transcend the profane, individual condition and to attain a transtemporal perspective."[4]

A PRINCE OF OLD EGYPT became increasingly depressed, we are told, at the condition of the displaced tribal peoples building store-cities for the Nile River cityculture. Why this prince, who may have crushed a disturbance in Ethiopia, should identify himself more and more with these miscellaneous, wretched people is open to conjecture. Was he a rebellious aristocrat seeking to form a partisan army? Did he already have a secret band among these slaves? Or, was he really somehow one of them?

That the Moses story is a patchwork of diverse, even conflicting traditions is generally agreed upon.[5] What interests us now is the alteration in Moses between his flight and his return. In this interval Moses experiences an epiphany, or moment of enlightenment. And it is presented in the classic tradition of all the ancient saints, seers, and wanderers in the unknown. The four steps of the wilderness experience are clearly depicted.

As the episode opens, Moses was living at Pharaoh's court, but was torn with some powerful inner conflict. His 'madness'—which somehow involved the authorities, the slaves, and himself—slowly overcame him. In a fit of anger he killed one of the harsh Egyptian work captains. One cannot help but wonder if this incident, somewhat out of character for a prince, might not cover an abortive rebellion and an attack on an Egyptian outpost. In terror for his life, Moses escaped, his familiar world in ruins.

Striking eastward into the mountain wilderness of Midian, Moses 'gave up' his persona as Egyptian prince and allowed himself to be what he could be, a wandering Bedouin shepherd in the extended family of a Midian priest. For years he lived in this unroyal setting, adopting the austere non-city desert life with its patient attendance and reliance on nature. Thoughts of the enslaved *Habiru*, for whom he could do nothing, still tormented his mind. And at long last a strange thing happened. While out with his flock his attention was caught by an anomaly, the so-called 'burning bush.' And he became *fascinated*:

> Moses noticed that, although the bush was on fire, it was not being burnt up; so he said to himself, 'I must go across to see this wonderful sight. Why does not the bush burn away?' When the LORD saw that Moses had turned aside to look, he called to him out of the bush, 'Moses, Moses.' And Moses answered, 'Yes, I am here.'[6]

The operative sequence in the Moses story from his initial flight down to the bush episode is that what seemed genuine was abandoned for what appeared to be nothing. Then, in real time, the values altogether reverse: what seemed nothing became all. Moses had totally given up his cityculture position and pretensions in the magnificent Egyptian civilization. But he thereby was freed to live in wilderness and to wonder at the unexpected—even to turn aside from the ordinary to investigate what seemed bizarre and frightening. As Moses approached the unfamiliar in a mind of inquiry—rather than in a knowing, recognizing, projecting state of mind—wilderness could give him his time-initiation and his prophet powers. The remainder of the cycle, we might say, is history.

Gautama Sakyamuni, the prince of a small Hindu kingdom, became deeply depressed, like Moses, at the fate of the common people on whose exploitation the kingdom depended. This prince slew no one, but he did finish off the image of a divine being which he had personified for his little world. Sickened at the incongruities of life, he disappeared from his palace in the middle of the night, abandoning his wife, child, and throne. He turned up in a community of other disaffected men living as best they could in caves deep in a forest. There, Gautama tried to outdo all others in the practice of hard austerities, fasting, going without sleep, and the like. Each of these self-inflicted discomforts was a rejection of life-as-usual and a powerful cry for the attention of the creation-spirit for meaning and insight. At last, Gautama reached a point of frustration not only with the culture he would escape, but especially with his attempts to *achieve* his way into an intimate relationship with creation by means of 'holy practices.' In short, he gave up even trying to give up.

This spontaneous collapse and disintegration of effort to 'find meaning' brought Gautama at last to the Bhodi Tree. There he sat immovable in total abandonment to the incomprehensibility of life. It was out of this shattered state of mind—which could locate no meaning in yesterday or in tomorrow, in traditions or in hopes—that Gautama received his flood of illumination. As the Buddha, or Enlightened One, he went on to Deer Park and held his famous conversations which gave a new turn to man's thoughts.

From the adventures of Elijah the Tishbite in 1st Kings we learn that the discovery process is not one moment of 'giving up' and retreat, but many of them. Elijah was already a prophet in Israel, the one who could call fire from heaven, and had many experiences within the intuitive mode. But when Jezebel, the Queen and sponsor of Baal worship (if not herself a Baal high priestess), cursed him, and vowed to kill him, he was overcome with terror and fled into the wilderness.

> He came upon a broom-bush, and sat under it and prayed for death: 'It is enough,' he said; 'now, LORD, take my life, for I am no better than my fathers before me.' He lay down under the bush and, while he slept, an angel touched him and said, 'Rise and eat.' He looked, there at his head was a cake baked on hot stones, and a pitcher of water. He ate and drank and lay down again. The angel of the LORD came again and touched him a second time, saying, 'Rise and eat; the journey is too much for you.' He rose and ate and drank and, sustained by this food, he went on for forty days and forty nights to Horeb, the mount of God. He entered a cave and there he spent the night.[7]

The cave was no less than an oracular cave. In such a cave one might experience dream-visions of the numinous. (As we know, the early European bison and mammoth hunters also resorted to these caves of mystery. And later, across the Mesopotamian delta, the temple mounts were man-made recreations of these natural abodes of the timeless.) Following all the signs of a divine approach—winds, earthquake, and lightning—Elijah was questioned and received his commission. What we have reproduced here was even then ancient beyond memory: the individual fleeing the order of a dominant culture and seeking in the not-known for a new, larger gestalt of reality.

> A great and strong wind came rending mountains and shattering rocks before him, but the LORD was not in the wind; and after the wind there was an earthquake, but the LORD was not in the earthquake; and after the earthquake fire, but the LORD was not in the fire; and after the fire a low murmuring sound. When Elijah heard it, he muffled his face in his cloak and went out and stood at the

entrance of the cave. Then there came a voice: 'Why are you here, Elijah?'[8]

In India the villagers accounted it a blessing to take sweetened rice cakes into the forest caves to feed their uncommon men, their yogis and saffron-garbed holy ones. Likewise, the Judeans, Galileans, and others in the Ancient Near East had their own extended tradition of desert prophets. Such men had long been called 'sons of god.' In centuries past they had wandered from time to time in out of the desert mountains, teaching, healing, prophesying. Some dressed in skins, their long hair unkempt, eyes ecstatic, words strange.[9] Like the yogis, they possessed nothing, but survived with other wilderness creatures on the bounty of the day—roots, berries, honey, and occasional gifts from those peasants who felt they were blessed through sharing with a divine being. And, even in the later, difficult days, it seemed that the wilderness might still provide a prophet, a man of vision and power.

What these unusual men—who lived scrupulously *with* creation and *in* its time—taught by means of didactic stories and enigmatic parables seemed puzzling even then. Many of the distraught villagers who were being Hellenized looked for a guerrilla leader, one recruiting a warrior-band. They hoped for a new David or a Judas Maccabaeus, who would throw out the foreign princes and reestablish the old sacral kingship. What they found in the man of wilderness was more often a kingly presence than a king—a charismatic healer with strange powers who would neither lead nor rule. A man of mystery, disinterested in action, skeptical of changing tomorrows. A teacher who spoke of man's place in creation as a son in his father's house.

JESUS EMERGING FROM THE GALILEAN wilderness was just such an extraordinary man. Many others, we know from the Qumran Scrolls and elsewhere, had, in the centuries before Jesus, carried on this special tradition of living and wandering outside whatever culture and kingship were dominant at the moment. Living in the remote hills, they were able to carry on ways of thinking that predated the times of trouble.

Palestine, the country and its people, had been ravaged almost continually for some 700 years by the armies of invading citycultures—Egyptian, Philistine, Assyrian, Syrian, Chaldean, Persian, Greek, and now Roman, each looting and bearing off tributes of precious metals, jewels, wool, timber, and slaves. Within recent memory the Jews had made a valiant effort under the Maccabees to throw off the Greek yoke. But in the end they fell to another, more powerful, more implacable kingship. The tension in men's hearts between the dreams of what life ought to be and the life they lived was excruciating.

Into this highly charged situation, from which the Jews could see no way out, there suddenly came an archetypal figure from the past. For centuries no prophets had arisen in this land to bring forth fresh words from the Lord. Prophecy as of old was not just discouraged by Jewish Law, but forbidden as blasphemous. The Age of Prophecy was considered past and closed. To step forward now and speak the word of the Lord in a voice of ecstasy was punishable by death.[10]

Out of the "wilderness" John the Baptist appeared on the banks of the Jordan River. As a holy man busy with his rituals at the edge of the sacred river, John seemed religious enough. But in his reinterpretation of time he was every bit the revolutionary:

> John wore a rough coat of camel's hair, with a leather belt around his waist, and his food was locusts and wild honey. People flocked to him from Jerusalem, from all Judea, and the whole Jordan valley, and were baptised by him in the River Jordan, confessing their sins.[11]

This rite of lustration by water, it is important to note, was not Jewish. From our recovery of ancient libraries we now know that baptism was a borrowed tradition. The breakaway Essenes in their camp by the Dead Sea did have large cisterns and made some ritual uses of water. But as a developed ceremony and sacrament in early

Jewish rites, baptism is not to be found. On the other hand, it did figure significantly in the rituals of neighboring tribes and kingdoms.[12]

Why then was baptism in the Jordan suddenly of such great importance? The answer may lie not so much in baptism *per se* but in the mystique of that particular river. The waters of the Jordan had already appeared on several occasions in Jewish lore. As the Mosaic priests carrying the Ark of the Covenant stepped into the Jordan, its waters had divided. The peoples who had come out of Egypt some 40 years before were able to cross without incident to the land 'promised' them by the Lord.[13] This parting of the river became in later centuries a manifestation of shamanistic powers much akin to calling fire down from the heavens. Elijah the Tishbite, "a hairy man with a leather apron round his waist," was wont to roll up his cloak and strike the river with it. And the waters of the Jordan would divide so that he might cross like the original children on 'dry ground.' Elijah's transport into the heavens was witnessed by his disciple, Elisha, who as a consequence inherited the robe and its shaman powers to affect the Jordan.[14] The river also figured prominently in the story of Naaman. This afflicted general bathed in its waters seven times as suggested by Elisha and found his leprosy miraculously cured.[15]

We generally think of baptism as a simple 'washing away of sins,' a relieving oneself of guilts and regrets, of serious transgressions of cultural taboos. In ancient Jewish tradition, however, the restoration of the individual to the good graces of the culture was affected otherwise by a variety of sacrificial offerings—burnt offerings, gift offerings, whole offerings, thank offerings. The community rid itself of collective sins by driving out a 'scapegoat.' (The original sacrifice was likely human.) In our own time baptism has evolved into an emotionally charged, all-in-one mystery experience of the death of our bankrupt personality and our rebirth as a new individual. This baptism-as-conversion is much akin to the mind altering procedures studied by William Sargant in his important book, *Battle for the Mind*.[16] For an already agitated person, uncertain of his worth, the emotion-drenched baptismal ceremony can provide the moment when his desperate psyche makes its paradoxical 'renewal.' Being thrust firmly under the water does suggest a real demise by drowning. And being pulled up from the waters—choking and coughing, emerging from the 'amniotic fluids' of the magic river, seeing the concerned, but happy faces of his new friends—could well give a disturbed individual the sense of integration into a loving, caring community.[17] (Here we might also compare the psychology of baptism to the psychology of the Najavo 'sings' which function to reconstitute the broken ties of the kindred.)[18]

But whether the baptism we know—admission to a moral community and acceptance by the god who oversees tomorrow—was the "Baptism of John" is something else. Although the Synoptic Gospels clearly state this baptism was a washing away of sins, there was curiously, as we noted, no precedent for it. (And the Jews, like all peasant people, were sticklers for maintaining traditions.) Further, 'sin' denotes not just injury to others, but, at a deeper level, the heresy of advocating what the community perceived as an erroneous definition of time. (Sin was being out of synchronization.)

Now, while baptism did not figure in Jewish history, it so happened that a water rite of another sort had been extremely important. Oppressed people everywhere resort to ambiguities and disguises. Simple words, gestures, activities are encoded with occult meanings known only to the underground. In this anomalous appearance of a Jewish baptism in the Jordan we have most likely the effort of a desperate people to turn back the clock. Going to the waters of the Jordan signified a renewal rite which had to be carried on covertly lest the Jerusalem priests and the Romans suspect.

The Hellenistic perception of time had already obliterated the Jewish language, corrupted the Jerusalem cult, and was tearing apart the Jewish home. The energetic Selucid leadership had urbanized Palestine. Ancient market villages were leveled and replaced with modern cities of gleaming, pillared forums, paved roadways, and spacious amphitheaters. Gymnasiums and baths were constructed and athletic sports introduced. Long before the Roman occupation, the Jews had already been pushed from peasant time to citytime. "The young men of the aristocracy went over with a will to the new point of view. They formed a guild of the Ephoboi and flaunted their Greek cloaks and broad-brimmed hats about the streets and in the temple area. To be an athlete became the craze. Even the priests caught the mania."[19] All evidences of Jewish authenticity and Jewish time were being lost. If the time of Moses were to be reenergized, the rites would have to be unusually moving.

At this critical juncture there suddenly appeared from the wilderness an ancient figure—the archetypal Hebraic prophet-shaman himself. Dressed in a hair coat and girded with a leather belt, this worker of miracles—the mouth and tongue of the Lord, the master of fire from the sky—had in effect stepped from history.[20] Such a man had long been forbidden under penalty of death from prophesying or shamanizing. So John came instead with his baptism and his news. The "whole Judean countryside" flocked into the wilderness and found him waiting at the banks of Jordan.

The scenes that followed were public theater of uncommon, electrifying power. One marvels at the exact fit of the ceremony to the need and the magic that resulted. Now, the mythic point in time at which the Hebraic community was born had been the original fording of the Jordan. The possession of the 'new' lands meant the beginning of a new time. In the figure of John these latter-day Israelites, dispossessed of their lands and their time-values, perceived, not just a herald out of the wilderness, but a heroic leader out of the Mosaic age. If Jesus saw in John the return of Israel's master shaman, none other than Elijah, we can feel sure the multitudes did likewise.[21] The individual's plunge into the waters of the

Jordan created a mystic identification through time with the ancestors out of Israel's glory who had also been washed by these same waters. (A river may 'flow,' but in another way its waters are always the same and, therefore, unchanging.) "John's Baptism" was a shrewd and profoundly affecting recreation of sacral time—or *illud tempus* to use an expression favored by Eliade.[22] John would 'force' the reappearance of the covenant community and the God of that covenant by reenacting the original time-magic. His efforts at the Jordan can be compared in ways with Venner's charge through London and with the Ghost Dance of the beleaguered Indians of our Far West who yearned for a return to the time before the white man.[23]

Of what thoughts Jesus had as he went with others down to the Jordan to see John (was he really Jesus' cousin and/or a confrere in some secret society?), we have been given scarcely a hint.[24] But that he joined the disaffected in seeking out John and participated in this restaging of Israel's origin time tells us much. We can imagine with some reason that he felt burdened like many others with the general hopelessness. However the Jews might scheme for escape from the pressure of Greco-Roman cityculture, their ability to *do* anything was nearly zero. An end of time had arrived for the Jews. Evidences of their past had been destroyed by successive foreign kingships. The future did not exist. Only an act of the Lord of time—the arrival of His Kingdom with a new meaning for the passage of days—could forestall the death of the Jewish soul. The "end to the appointed time," as foretold in Daniel, 11:28-35, seemed at hand.

The timing and staging of this unusual event at the Jordan were brilliant. The Baptism of John invoked the Mosaic-Davidic order of time. This very river had been witness to the sacral origins of the Hebrew people. The immersion was a revival of the great beginnings, washing away all the intervening blemishes of foreign, or profane time. The Baptism reenacted the historic river crossing by which the covenant people entered the new Eden, or "Promised Land."

What more intense political-religious reenergizing could these splintered people experience than returning to such a climactic moment? Feeling themselves the new embodiment of those "children of Israel," Jesus and many others dipped into the waters of the Jordan and reexperienced the magical time of entry. They emerged transformed—once again they were Mosaic men and women. By this ritual in the Jordan, they had rededicated themselves to the continued integrity of their sacral community and its concept of the "Day of the Lord"—that near-future when wrongs would be righted and sacral time reinstated. In such an emotion-charged scene, perhaps accompanied by wails, chants, and the ululation of the women, an ecstatic experience of time, as undiminished as the holy river was everflowing and ever the same, could well have occurred:

> After baptism Jesus came up out of the water at once, and at that moment heaven opened; he saw the Spirit of God descending like a dove to alight upon him; and a voice from heaven was heard saying, 'This is my Son, my Beloved, on whom my favour rests.'[25]

One might reasonably expect from this passage that Jesus would begin his ministry right then—that this blessing had endowed him with his powers as a prophet-savior. But not so. From here his path lay through the uncharted. This baptism became for Jesus that moment of destabilization which is initiatory to what we are calling 'the wilderness experience.' Being seized with the explosive idea that he himself might be the long awaited "Suffering Servant" foretold by 2nd Isaiah must have come as a hammer blow.[26] To discover oneself a prophet in a world that killed prophets was surely as disorienting as was Jezebel's vow to slay Elijah. The dove and Voice experience was for Jesus the threshold not yet of ministry, but of his own search for time and its meaning.

The experience at the river was advertised as cleansing one of profane time. Covertly, it was a reaffirmation of the tribe's own sacral time. But it also carried a still deeper implication. The flow of a river, as we have suggested, is analogous to the 'flow' of time. No fixed points can be engraved on water. In a like way you cannot pinpoint with your dividers where the past leaves off and the future arrives. (Saint Augustine in his celebrated Book XI of *The Confessions* tried mightily.) Both time and the river have this wonderful illusion of motion. But seen again, they move not. In the mythopoetic world of the Ancient Near East the 'river of time' became manifest in the Jordan. To cross this river was man's immediate encounter with time. And in the shaman's ability to halt the flow and divide the waters with his 'cloak' was concealed a great mystery. His power came from his discovery that time could be seen in various ways, one of which caused its apparent movement to stop. And in this stoppage there was revealed a narrow corridor between the walls of yesterday and tomorrow. We have here one more example, then, of the *difficult passage* so often mentioned in the legends of the questing heroes. In his crisis at the river, or soon afterwards, Jesus may well have had his initial insights into the illusion of time as flow.

> Jesus was then led away by the Spirit into the wilderness, to be tempted by the devil. For forty days and nights he fasted, and at the end of them he was famished.[27]

Like others whose cultural sphere has been shattered, Jesus turned to the unknown and became a wanderer in the timeless. He had been 'initiated' into the natural condition of creation, which is the unmarked and uncertain. No longer can man be the determiner; he becomes instead the listener. This transformation, when permitted, will not be from a condition of order to mere chaos. Instead, the mind disgorged of cultural limits will find its way (the 'difficult passage' in a state of timelessness) to reconstitute the dialogue that early man had with the unknowns of creation.

This spiraling around of the mind to capture anew what was lost in the intervening 10,000 years of cultural adaptions, we are referring to as the intuitive, or wilderness mind. It cannot be quite the same as the hunter's experience with the unmanifest. Living in real time had been, for the hunter, a long tradition in itself. He did not think of and fear wilderness as we of the city do. Nor did he have in his thinking the hostilities and alienations of the peasant mind and the kingship mind. To unriddle those alien thought modes and to recover a sense of real time—while yet in a hostile environment—creates an alert mind similar to the hunter's, but with a strong addition of skepticism—altogether a *4th mode.*

But as with early man, we shall see again and again that an experience of disorientation—whether a simple aimlessness or an appearance of schizophrenia, of coma and trance, or a wandering unkempt, hungry, and sick (the 'real world' of tradition and ambition being completely rejected)—is a prerequisite, as chrysalis to butterfly, to the emergence of this 4th-mode identity.[28] Not all wilderness wanderers are going to return to us with a new order of insights into the meaning of events and life. Nor will all appear as extremities of the starving and disheveled. But not too far behind every discovery person will lie periods, short or extended, of wandering, of being thoroughly distracted, of being 'lost.'

Jesus, on discovering himself to be the "Suffering Servant," fasted himself into a hunger narcosis. A series of transphysical experiences with the Tempter followed. 'Possession by the Spirit' has often been used to describe a sense of transcending ordinary, secular, time-as-motion. With this new feeling of invincibility and participating in sacral time—here, the divine tomorrow, or Day of the Lord—comes an inflation of power, the Tempter. The achiever side of the mind, always ready to fashion new tomorrows, is ever urging each son of man to dabble in the clay of the earth and affect destinies. By turning from both tradition (his family ties and the religious orthodoxy) and kingship achieverism, Jesus demonstrated that his mind had 'crossed the Jordan' into still another mode of time. As a knower of wilderness and the timeless, he was possessed by the intuitive. And he moved and spoke accordingly. The first words of his ministry were of time: "The time has come: the kingdom of God is upon you; repent, and believe the Gospel."[29] What was there to repent? The false god of Hellenism, which was cityculture and the triumph of tomorrow. What Gospel was there to believe? That time as flow was an illusion. The master of epiphanic time, Elijah, had reappeared. Iron Age tomorrows would no longer come about for the Hellenes or for the Romans. Nor were the glories of a Bronze Age yesterday to be sought by the Jews. The magic of the Jordan was reinvoked. And, wonder of wonders, a way opened through the difficult world.

> Then Jesus, armed with the power of the Spirit, returned to Galilee; and reports about him spread through the whole countryside. He taught in their synagogues and all men sang his praises.... There was a general stir of admiration; they were surprised that words of such grace should fall from his lips. 'Is not this Joseph's son?' they asked.[30]

Soon, however, Jesus spoke the harsh words of a prophet: the people of Israel did not deserve "the Lord's favour." Then these same villagers became hostile and dangerous:

> At these words the whole congregation were infuriated. They leapt up, threw him out of the town, and took him to the brow of the hill on which it was built, meaning to hurl him over the edge. But he walked straight through them all, and went away.[31]

Jesus and others like him had sought wilderness and reappeared strange and imbued with powers. Their experience of psychic metamorphosis and a mind independent of cultural validation was an ancient tradition in itself. How each expressed himself and challenged the conventional world was highly individualized. Strangely enough, however, the words and images that many of these wilderness-wanderers used formed a kind of timeless language. Its grammar was saturated with the present, with the interrelatedness of life, and with direct perceptions of what was little known. Each spoke of living within the world, not as a contestant with the earth (as depicted in Genesis 3:17-24), nor even as one of its creatures, but as a ray of creation itself.

> Look at the birds of the air; they do not sow and reap and store in barns, yet your heavenly Father feeds them. You are worth more than the birds! Is there a man of you who by anxious thought can add a foot to his height? And why be anxious about clothes? Consider how the lilies grow in the fields; they do not work, they do not spin; and yet, I tell you, even Solomon in all his splendour was not attired like one of these.... Set your mind on God's kingdom and his justice before everything else, and all the rest will come to you as well. So do not be anxious about tomorrow; tomorrow will look after itself. Each day has troubles enough of its own.[32]

"Is there a man of you who by anxious thought can add a foot to his height?... Consider how the lilies grow in the fields.... Tomorrow will look after itself." Few passages in Scriptures have created the unease in our theologians as have these simple direct admonitions and others like them in the Sermon on the Mount. All else attributed to Jesus—including his virgin birth, his altering the cosmic order by sacrificing his body for our 'sins,' his bringing a dead man to life, and other difficult wonders—can easily and readily be accepted by our divines. But Jesus's denial of yesterday and tomorrow (a theme of the wilderness-validated mind) is a most difficult and baffling concept for the citymind to grasp. A modern theologian, Martin Dibelius, in his book, *Sermon on the Mount*, makes an observation that is widely held:

> If we were to extend the content of the Sermon on the Mount so as to cover the whole of life...we would construct a revolutionary law, the fulfillment of which would throw the whole world into turmoil. But it was not Jesus' purpose to give such a law to the world of his day. * * * The deeds and words of Jesus were *signs of the Kingdom of God*, nothing more and nothing less.... His miracles are a revelation of heavenly forces and a proclamation of what God is going to do. * * * When we interpret the sayings as signs we

> can understand and in some way reconcile the apparent contradiction which was till now implied in our consideration.[33]

What God is going to do! The Jesus message was rather that God was not going to do anything that He was not already doing. This son of wilderness proclaimed, again and again, yesterday and tomorrow as irrational. "The Kingdom of God is within you," is how he affirmed the reality of the timeless-present from which man can never really separate, or alienate himself.[34] The 'Kingdom' became a metaphor which Jesus used to suggest the powerful faculty all men had, but rarely allowed to surface. The mind in its mode of 'neither yesterday nor tomorrow' was capable of perceiving truths and realities greater than the everyday culture and its hidebound traditions permitted. Men did not have to live as vassals in cityculture's tomorrow. "Render unto Caesar what is Caesar's and unto God what is God's."[35] Nor did he have to be terrorized by the past: "Let the dead bury the dead."[36] The new experience of time which came upon him in the wilderness revealed that creation had "many mansions" for man to explore. "Seek and you shall find" was an admonition to quest beyond culture's traditions for a discovery of that time which was not yesterday, not tomorrow.[37]

All this business of no past and no future sounds absurd to our ears. And it is quite understandable why Dibelius, and many others beside him, have tried to rationalize a way around such stumbling blocks as "Look at the birds of the air" and so forth. The whole idea of not caring for the morrow is, in the thinking of cityculture, perfectly outrageous.

Reality and power for the achiever-minded have nothing whatever to do with birds or lilies. Nor is the 'sane' cityman going to be caught waiting out anywhere in the creation's wilderness for his food. He is going to wring not only his dinner out of wilderness, but much more besides. Cityman has reasoned and justified his separation from the matrix of creation with lawyerly rhetoric.

Citythink's rubric here: *Life is without meaning, but man choosing his goals can give it meaning.* Cityman endlessly reworks and reshapes his environment, tearing down, building up, and tearing down again. He plays the achiever game of trash yesterday, mortgage the present, mail-away for a beautiful tomorrow.

Today we are much given to feasibility studies and windows of opportunity. To be is to be able to alter the future. Non-achiever thoughts cast in non-achiever language have little effect on our minds. Yet, with those charismatic images out of the Bible, the *Bhagavad-Gita*, and elsewhere, which the uncommon men of earlier ages have left of themselves, we cannot help but feel affected. We keep their remembrance as talismans, but hardly for their radical notions of time and mind transformation. Rather we resort to them as intercessors with creation when our plans for tomorrow now and then go awry. Or, at the least, we recall them as comforting, familiar icons, museum pieces, holy artifacts behind glass. Each, as an idealized anthropomorphic god, makes a difficult world seem right for many. But if we consider them as flesh and blood individuals set in an historical context, they can appear hopelessly disengaged from reality. After all, did the money changers stay out of the Temple? Or, did they move back in the next day? For most of us these wonderful, special persons serve little beyond being idealized social models in the achiever's 'real world.'

Amos left a few choice words for the achiever-kingship mind:

> It will be as when a man runs from a lion,
> 	and a bear meets him,
> or turns into a house and leans his hand on the wall,
> 	and a snake bites him.[38]

The fully alive man—the uncommon man—is just as puzzled as the achiever by what seems creation's unstated intent. He listens, however, for hints of what he has never before heard. He scans for what has not been seen. He seeks the unexplained behind the ordinary.

PSYCHIC METAMORPHOSIS

METAMORPHOSIS IS NOT A NEW, uncommon idea, even though it is seldom a part of our common speech today. The ancients were much taken with the way nature turns one creature into another. Ovid wrote a book which dealt with various examples of this magical change as described in that enigmatic folk-psychology we call myth. Cityman, however, does not like to think of himself as the ongoing product of unseen forces; we are individuals in command of our own changes. Cityman maintains he has defeated metamorphosis; we make every effort to prolong youth into middle age.

Nevertheless, many of us in our biology class wrote a small paper on how the tadpole becomes a frog. If not that, then how the silkworm by specific stages transforms into a moth. Once we received our grade, that marvel of nature's process—image merging into image—soon dimmed. It became another of those 'curious facts' which one learns about as a child, along with Indian lore, ghost stories, and folk ballads. From that time-order which occurs second to second in the organic world, we passed on to learn the important 'realities' of tomorrow. Achieverism—our destiny as man to bring cityculture values to a nature no longer curious but now seen as random and dangerous—is the 'name of the game' in the grown-up world. Cityman becomes adult, he believes, when he is capable of supplanting creation's process with his own linear determinations.

Metamorphosis has everything to do with a life-phenomenon reaching its fullest expression—its own peculiar ongoing maturity. Even man cannot escape the dance and cyclic change of images. If he tries, he damages himself. The pressure for this mysterious, astonishing movement through seasons and stages—each an image entire in itself, unmindful of what will follow, but resolving of itself by orderly phases into the next—is not a mystery given to some creatures and withheld from others. As far as we presently know, no creature springs into its maturity full blown. Each has its own special metamorphic pathway. The DNA spirals and many glandular secretions, we have found, play their important parts to ring in the physical changes which bring about what seems essentially changeless. But of psychic metamorphosis we are far from knowledgeable.

Out of all creation man is the only life-expression which struggles to thwart or ignore the metamorphic cycles—both physical and psychic—as they try to work their magic in him. He insists on knowing beforehand what changes may follow. He likewise demands the 'right' to cancel any upcoming change that does not meet his specifications. The organic push towards maturity through serial alterations comes from sources that lie deep in the man's nature. Maturity here means the full integrity of the organism, a development well beyond whatever it could possibly visualize for itself in an idealized tomorrow. Man's symbol-processing centers—concerned exclusively, as they are, with what may or may not be recognized and how the environment shall be used—are totally inadequate for guiding the maturity of so complex and protean a function as a man.

Metamorphosis means change without control over the direction of the change. Since this is the game that life plays, some creatures instill in their offspring a sense of how to approach the unexpected. In the world of the hunt, both parties are inclined to alertness—the very moment can never be fully anticipated. The natural creature learns early to face the not-known and relate directly, wherever it can, to what is beyond the recognizable. Out of such a challenge and response, or agitation and attention, come actions and conceptualizations not possible in any other way. But the unnatural creature, cityman, would thwart the built-in evolutionary, metamorphic compulsions.

Now, the thesis which cityculture advances for man argues that creation provides him with a body, but it is cityculture that provides him with his mind. This notion proceeds on the assumption that the cyclic process by which we are formed is finished when we reach the age of 18 or 21. Our bodies are then complete and we are pretty much on our own until exhaustion of the biological machinery some 50 to 60 years later. The decades intervening are ours to make of them what we will through 'free choice.' Admittedly, the body comes to its maturity on its own, without external help other than food and safety, through a process of images merging into further images. But man's mind, the city thesis holds, is independent of such an autonomous process. The individual has little to say over the dimension and contours of his body. The activities of his mind, however, he thinks are all in his control: the right to image and to play with images are the property of the individual to do with as he wills—or as the culture may convince him.

There can be no denial that today our common mind is supplied by cityculture. To some degree we are all tomorrow-driven. It is to cityspeak we must resort when we learn a trade, build our houses, and pay our bills. But being saddled with an achiever mind-set early on does not end the matter there. Neither does the fact that one must relate now and then to the achiever world and its skewed sense of time.

We must discover whether the metamorphic process—images periodically merging into further images—is really exhausted and through with the individual when he reaches 21? Or, does it have further intentions, in other areas, to go on transforming and particularizing him in special ways throughout his entire life? The question we now ask is whether this mind of inquiry—which predates the peasant-tradition mind and the cityculture-achiever mind—is not in some way an expression of creation's metamorphic process. Our physical bodies are subject to a growth by stages. Does our natural, acultural mind also 'grow' and realize its powers through stages? We are not speaking here of the age-related phases in the life of the youth and adult which have been much discussed in popular psychologies of late. What concerns us is how man deals with his immediate reality, moment to moment, through the agency of his mind. Are the energies which man represents going through some sort of metamorphic process?

The achiever relates not to the given world, but to an alien realm which is the fiction he has concocted and calls "the better tomorrow." His interaction with that hypostatized reality takes the form of a four-phased cycle which starts and ends in anxiety and repeats itself over and over. Each swing around tends to diminish the power of the individual until an early exhaustion is reached. Instead of a metamorphosis of growth we have a premature aging and diminution of faculties.

The wilderness mind, on the other hand, relates with a world different from the achiever's. Where the later would use the present as a resource to fabricate and make real his unreality, the wilderness mind would extract from the present as much intelligence as will be vouchsafed to him. The achiever, by destroying the present, effectively divorces himself from that situation through which larger intelligences might communicate with him. The bardic mind leaves the present relatively untouched, ready for further accesses of intelligence. In this relationship, as we have described it, his mind experiences a four-phased 'thought' sequence similar to the achiever cycle. Starting in uncertainty-wonder (instead of anxiety), it passes through search, exhaustion, illumination, and ends again in uncertainty-wonder.

As we learn the showbiz of post-industrial achieverism, each of us remains nonetheless that which cannot be denied—a process which is trying to express its own particular maturity. This is the covert, ongoing, unfinished business that no one has told us about and which cityculture steadily, consistently, attempts to defeat. As a result we are each a kind of psychic 'Baby Fae.'[1] Much as our immune system knows to reject a foreign implant, our integrity as an organism reacts to the intrusion of an alien time-set such as the achiever mind. Nonetheless, as we "come of age" powerful anxieties, aversions, and promises of distant rewards suppress our resistance. Thus, the field of our mind—which normally should be ever alert and nimble, ready to detect and question—becomes filled with a toxic sludge of cynicism, denial, and antipathy. By this inner turmoil our brain-mind is scarred; our life, narrowed and shortened. We fall victim to a psychosocial multiple sclerosis. We are everyone of us casualties of citytime.

That maturity which should arise in us naturally through our own highly individualized psychic metamorphosis (interaction with the unfamiliar) is routinely supplanted by cityculture's own process (interaction with the familiar). Already embedded in the citymind issued to us in young adulthood are the aversions which will abort efforts to probe past its restricted architecture and into what might have been our own real nature.

Cityculture detests the idea of cycles and argues convincingly that the linear path is the only way for individuals to get things done. Set goals, gather materials, assemble a work force, and produce the designated better tomorrow. Cityculture has been very successful in convincing us that anything cyclic means the same as standing still.

COULD IT BE THAT PERIODICITY—that seasonal resurging of natural energies, which has been around for a few million years and more—is not so wrong-headed a notion or as old-hat as we were led to believe? The achiever claims his trajectory towards his goal follows a linear pathway. We have shown this is only true if the phenomenon is viewed for a brief moment—as a snapshot in time. In reality, the other experiences of the achiever life, the despair of a goal, the rejection of the goal, the desperation to substitute a 'better goal,' and so on, make up the "remainder" of the four-phased cycle which is ordinarily disregarded.

While the apologists claim vociferously that whatever is cyclic is 'dead in the water,' without direction, such 'know-nothingism' is merely cityculture's defense. The fact, everywhere observable, is that seasonal activities in the natural world produce fantastic increases. Every cyclic process, if unimpeded and undistorted, results in a surplus of energy or in some other increase of nature. The rule here is: *Power arises from the periodic.*

Prometheus brought man the knowledge of how to make fire. The fire wheel—a rapidly spinning shaft of hard wood pressed into a base of soft wood—epitomized for the ancients that profound mystery which they saw all about them. Everywhere increases of energy, form, perception, and kinship arose continuously out of activity which ever moved in large and small periods. The great sage who came once an epoch to redefine the nature of time was known as the *chakravartin*, the one who turns the wheel of dharma.[2]

The whole cosmos, to be sure, was perceived as one vast astrologic turning, wheels within wheels, cycles within cycles—infinite microcosms within infinite macrocosms. (And all this long before atomic physics!) Even the woman who spun thread with her weighted whorl was thought to be reproducing the universal motion of the heavens around the pole star. Penelope weaving by day and unraveling by night was a representation of the mystery of resurgence. Perhaps the classic depiction of the cosmos of four directions in its cyclic motion was the striking image of the *merkaba*, or Throne of God, in the first chapters of Ezekiel:

> As I looked at the living creatures, I saw wheels on the ground, one beside each of the four. The wheels sparkled like topaz, and they were all alike: in form and working they were like a wheel inside a wheel, and when they moved in any of the four directions they never swerved in their course. * * * When one moved, the other moved; when the one halted, the other halted; when the creatures rose from the ground, the wheels rose together with them, for the spirit of the creatures was in the wheels.[3]
>
> The sound of the wings of the cherubim could be heard as far as the outer court, as loud as if God Almighty were speaking. Then he told the man dressed in linen to take fire

> from between the circling wheels and among the cherubim; the man came and stood by a wheel, and a cherub from among the cherubim put its hand into the fire that lay among them, and, taking some fire, gave it to the man dressed in linen; and he received it and went out.[4]

Fire from between circling wheels! A dynamo spins and the 'electric juice' produced by coils passing through magnetic periods lights a city and energizes a hundred factories. The cyclic was long ago perceived by the pre-achiever mind to be creation's way of releasing its stupendous flows of energy and wealth—or *power.* When the cyclic was interrupted or ceased altogether, there was no warmth, no increase, no life. And it is still true today that without the mystery of the cyclic in him no creature, not even cityman, can long survive. The persuasion that the cyclic is a condition of going nowhere is an argument that should be radically challenged. And one should also question who is advancing such an objection and why.

The achiever mind is poorly prepared for the cyclic, or metamorphic idea. It may be all right to speak of frogs and butterflies and fields of corn as cycling through neatly described phases. But man, we achievers insist, is quite different. To be sure, we all know that his body experiences identifiable periods of growth. Piaget, Erikson, Gesell, and others have also argued that the individual as child and young person develops through emotional and psychological stages which have characteristics that can be defined and dated. In a like manner Jung held that man finds his "individuation" through a serial discovery of archetypes and marriage of opposites. In her bestselling book, *Passages,* Gail Sheehy described the extension into adulthood of the idea of life-as-stages. For most of us, however, the thought of our adult lives being controlled by unseen forces is upsetting. Very simply, this conception of living by stages and thresholds is counter-achiever. It disparages the individual's power of free choice and, for many of us, smacks of the medieval.

The idea of "right to choose" is a power notion as intoxicating as opium, hypnotic as a mirage. Before our delighted, hungry eyes we see visions of the "promised tomorrow." Only a little effort, it seems, will put us there. Once infected with this conception that tomorrow can be so much clay in our hands, we feel 'on course' and 'grown-up.' Maturity for cityman arrives, we believe, when the individual makes the transition from the resource being processed, to the processor and exploiter of resources. The adult is he (or she) who can resist being shaped and can actively plan and reorder his (or her) environment. In our minds we have created this indefinite, expandable alien world—exterior to creation and its cycles—which we call tomorrow. There, we believe, it is possible for us to become the master of changes.

For all of its 'can-do-ism,' however, the achiever mind is incapable of the one basic act which has been most fundamental to man-the-creature. It ranks alongside and at times ahead of hunger, procreation, warmth, and the like. That function is the ability to admit to not knowing, to face and 'behold' the unknown—that is, to allow its existence and its effects. The notions of recognition and use may be handy in the worlds of yesterday and of tomorrow. But in the world of *real time* they will not do. The 'stick' on the ground recognized as a stick may in reality be a cobra.

The 'primitive' saw the world as ever in subtle movement and observed closely to discern unfamiliar, curious variations of the moment. Such naturally occurring differences were more significant to him than a 'resource' recognized and assigned a use. The former was genuine; the later, artificial. In no way can the subject of natural change, or metamorphic process—be it butterfly, tree, or man—know beforehand what the next stage of transformation will make of him. Here, we can have neither recognition nor use. How much does the tadpole know about becoming a frog? The process that it represents has its specific cycle and order. The potential frog does not have to *think* itself into being a frog.

IN THE PROCESS OF PSYCHIC METAMORPHOSIS, which man's mind represents, each successive phase has its own separate lifetime: it demands its own 'time of troubles,' its own discovery problems, and its own maturity before it will release the trace signals which will trigger birth into the next stage. Even within a period things refuse to remain orderly. But as one image blurs into the next, surprise and uncertainty are everywhere. One may try to control this 'disorder' by setting goals, but they tend to drift like balloons from the hands of a youngster. If we do succeed in arriving at a few self-visualized targets, we find they lack the magic to excite us and become quickly disillusioned with them. Whatever actions we attempt in our profound uncertainty may seem futile at the start, but we often continue solely for the sake of being occupied and in motion.

As each phase of the metamorphic experience nears its completion, what seemed unknown, even threatening, becomes through the working of this 'wilderness process'—by a kind of magic almost—significant and purposeful. With the influx of insight—the unexpected "fresh idea"—what was confused and unfamiliar falls into a wholly new order. All words and relationships are redefined. That which moments before we had been unaware of, except for its disturbance of us, stands revealed. This enlargement of understanding—through *allowing the disturbance*—becomes for the wilderness mind its own 'increase of nature.' Such an epiphanic instant—a glimpse of the unknown and its inclusion—some have referred to as a "moment of truth."

The cusp between seasons is short. At the point of discovery one cannot linger. The metamorphic cycle means movement, but not an achiever movement towards a recognizable tomorrow. Each created organism would be marched steadily through many unfamiliarities towards a resolution of the not-known (or "big question") which is the sticking point of each position. The euphoria that comes when the chaos of the moment seems to explain itself is the maturity (or 'increase of nature') of that stage.

No sooner has it occurred, however, than one is plunged into the next period of unfamiliarity and its chaos. This ongoing experiencing of the not-known by stages, first as disturbance, then as a new and larger order, is the process of psychic metamorphosis.

The mind of man is not all blind, angry forces clashing in the dark. Nor is it the plaything of appetites trying to appease themselves with delicious new tomorrows. The special power of man's mind is its faculty to encounter unknowns waiting to be discovered and to pay close attention to them. His mind needs this arousing by the unfamiliar just as a plant needs soil and sunlight to expand and grow. Nietzsche wrote perceptively, "One must still have chaos in oneself to be able to give birth to a dancing star."[5]

Formerly we believed our mind was composed equally of memory and choice. But these are cityculture categories: memory is time past; choice is time tomorrow. The real nature of our psyche is otherwise. The experience which we call our mind is nothing if not the eye by which the unknown becomes revealed, before which wilderness plays out its intelligence. And if we are concerned moment to moment with our choices for a projected tomorrow and guided by recall of words of authority, we shall not experience the mind of man in its true function—the detector of the shadows and rustles of the not-known.

To attempt to substitute scenarios and desired end-use goals, for an ongoing metamorphosis is to interrupt the living, possibly to kill it. In the true wilderness experience, therefore, one is not goal-oriented, but *process-oriented.* Here, every man is a question, his own question. It sounds strange, but one must allow the chaos of the undefined aroused by his inquiry to happen to him. (Not as a victim gives in, but as the explorer discards an unreliable map and sets about making his own trail marks and readings of his position.) Otherwise, without entertaining the unrecognizable, there will be no phase maturity. One has to 'move forward,' almost blindly, with more trust to nature and her ways than one would like. One moves not to make things happen, but to gain new perspectives on one's sector of wilderness.

The sense of achieverism is so deeply ingrained in us that it is repugnant for us to 'let go.' Our whole nature has been trained to impress onto the environment our limited mental images of what shall be familiar. In wilderness, however, the situation is reversed. It is the unknown that would step by step impress its patterns and reality on man's mind. Learning to float on your back was a similar letting go. The buoyancy of the body, however small, is always enough to keep one's face above water—if we relax, that is, and permit it to establish its own equilibrium. Letting go to the not-known, however much the achiever objects, is not letting go to blankness or 'nothing.' Quite the contrary. Any not-known which is capable of disturbing us has to have a greater reality than the busyness with which we were occupied. We can project "great ideas" into tomorrow for our entire life and miss out on the significant experiences that were at hand all along.

This being lost in a wilderness of any kind is more than a matter of discomfort to the achiever. His mind, being severely limited to the confines of the recognizable, is terrified at the prospect of losing hold of tomorrow. He may experience hysteria, hyperventilation, and other panic symptoms. Running from a lion, cityman meets a bear. Exhorting the achiever to release his death grip on the future by his will power alone sounds feasible, but will fail. The achiever cycle is similar to a gill net: it works by illusion. The fish thinks he can slide through, but is snagged nonetheless. Even the idea that 'letting go' on purpose might somehow produce a more effective life-style and life-product is in itself a goal. And quicker than the flick of a bat's wing one is swiveled 180 degrees back into the achiever orbit.

THE CROSSOVER POINT

THE SELF OF CITYMAN IS FRAGILE. Based entirely on making choices within an environment of commodity and service menus, it lacks the toughness and adaptability of the hunter and even of the peasant. They were both accustomed to improvising, the hunter especially. But the achiever is most vulnerable when the specialized pathways of the city break down and leave him on his own in real time. It is then, and only then, that he may be motivated sufficiently to question the meanings of choice, tomorrow, and the mind he has inherited.

The very act of choosing throws one into tomorrow. The goal selected as a familiar to be realized is not in the present, never in hand. It will very likely be visible (if only in the mind's eye); it will certainly be something tangible, previously known, and imminently *recognizable*. The subject of the choice, as a consequence, can only be sensorial. Being of tomorrow and sensorial, the goal falls into that field of control long established and governed by cityculture—what the culture recognizes as tangible exists; what it decides to deny will not be seen at all or only dimly as an illusion or fantasy. That self which engages in sorting out and evaluating the familiar is the border-self we call our *conscious mind*. Such a mind, which deals in sensorial experiences and in the choices between them, is in reality not *our mind*, but the city's fiefdom. It is the alien in our psyche.

The harmonious functioning of the non-wilderness mind—whether peasant or cityachiever—is based on two factors. There must be a ready access to memory. And the individual will have given that memory the power of attorney to make choices for him. The peasant-memory programs role models, from everyday to heroic, and positions them in a spellbinding calender of gorgeous sacral events. The memory of the cityachiever rejoins segments of the past into transmogrified happenings called tomorrow. In either case our so-called thinking becomes a rapid sort of what we already know. "Before there is knowledge, there must be memory," is the lead sentence in a *Newsweek* cover story on mind and memory.[1] The whole of our mentation would seem to be involved with memory. As a consequence, memory in the non-wilderness mind takes on a life of its own, determining selections and charting tomorrows.

Wherever our ability to choose, however, is blocked for one reason or another, memory (which is our kit of workable choices) experiences a denial. A struggle ensues. When further choices fail, this collapse of our system of choices triggers a memory death. The problem becomes not one of ransacking further for some better choices, but of going beyond memory and 'feeling out' the contours of the unfamiliar.

Moments come in the life of every achiever when the linear breaks down. The complex efforts to reach the goal end in disaster. The goal attained proves worthless. The pleasure purchased was hideously boring. The expensive vacation was not where one really wanted to be. Every achiever knows the dead end, the brick wall, with some intimacy. With nowhere to go and nothing to try, 'thinking' fails. Such exhaustion, though painful, can also be fortunate. When the individual's linear imaging is forced to a full screeching halt, his own natural mind will have been uncovered. Briefly, it lies exposed, free of the culture's bias and limits. In this fertile condition, the mind is impregnable by unsuspected, surprising images of a larger scope. At such an opportune moment the wonder-game we experienced all too briefly in our childhood has an opportunity to start again.

Instead of a memory of many workable choices, what is required of us is that ongoing, enlarging experience known as *perception.* The peculiar and wonderful potentials of man's integral mind may only reveal themselves to us when the power of memory to arrogate our psychic energies falters and perishes outright. With the death of memory as operant consciousness—here, not memory as simple recorder of familiar events, but as *master* of events—the metamorphic process may again work its unconscious magic and open the way for an experience of the unexpected. If the individual suffering a breakdown of the usual cannot find his way to experiencing the rich life beyond memory, beyond picking and selecting, his mind may appear to cave in altogether.

Although the achiever mind and the discovery mind seem poles apart and unrelated, they do have one phase in common. The four stages of the *achiever cycle,* as we have described it, are: anxiety; search for and determination of goal; progress towards goal; achievement of and trashing of goal (which reenergizes anxiety). The four stages of the *discovery cycle* are: collapse of familiarity; heroic efforts to reestablish the usual fail; wandering in wilderness; experience of fresh perceptions. Attempts to communicate insights reveal greater complexity and a host of new, troubling uncertainties.

The areas shared between these two cycles are the first stages—what we are calling the anxiety phase in the achiever cycle and the collapse of the familiar in the discovery cycle. Essentially, the triggering situation for both mind-cycles is the same. The unexpected has happened and the flow of events has been derailed. Choices are no longer available. Expectations are canceled. The appearance of time as a rational sequence of events has exploded. In short, the mind has been pulled out of artificial time—either the past or the future—and dropped into real time. In this state of suspension, the achiever may continue for a while, as a nervous reflex, to reach for the last straw of a chance. The wilderness man, however, looks about for hints of that indistinct, larger situation he has 'bumped' into. The initial loss of familiarity was similar, but the diverse reactions of the two minds are significant.

The healthy individual will be brought to full attention by a bump, a loud noise or a flash of bright light. Psychologists call this reaction the "orienting reflex." Even babies only a few hours old will pause in their sucking as a reaction to noise, or flash, or touch of a soft brush on its cheek. The reflex is the organism's innate response to its environment. The smashup of one's cherished plan (i.e., time alienated as tomorrow)— much as the sharp sound and the flash—is another such signal from the environment to pay close attention.

The achiever will investigate the strange noise or light until satisfied; he will lock the swinging shutter, tighten the dripping faucet, turn off the TV. But his mistaken apprehensions of time—even with his egregious constructs and alienations lying in ruins all about him—he will not question radically. Instead he will puzzle for days to see where the details of this or that scenario might have gone wrong. His hearing and eyesight are reasonably good—both the sound and visual disturbances catch his attention. But his orienting reflex fails when his time-constructs collapse. That sensitivity to real time with which he was born has gone dead. He has no conception of that which might be trying to get through to him. Still bent on tomorrow, the achiever circles and thrashes about—much like a blind man beating with his cane, lost in the middle of a highway. He searches for a new goal or for ways to redeem the failed one. He asks others for advice and direction. Friends, counselors, priests, and ministers are importuned. Having a job may help ease his time-crisis, but even here he may go from one to another. If his difficulty in fixing on a goal to absorb his energies continues, he may turn to ways of dulling both his pain and his sense of 'alienation' from the achiever culture.

The man of wilderness, when his schedules run into trouble, accepts the orienting reflex. He will pause in a questioning, watchful mode—even as the hunter would scan the forest for subtle changes. A meeting he planned long in advance goes awry. A piece of work he spent months completing is rejected. As he is driving along, a car backs out of a driveway and smashes into him. He sends others on an errand with carefully written instructions. When they return, nothing was done right. He tries to explain his findings to others, but discovers he has not yet completed his own understanding. Expectations collapse. The man of wilderness looks for what is happening outside his peripheral vision. Is there some larger event of which he is not aware?—and for which the definitions have not yet surfaced?

Both minds, the achiever and the discovery, find themselves now and then torn from artificial time and set down in real time. It seems the human thing to 'fill' that time with activity, to get into motion, to work towards a new plateau. Seen another way, however, this is not a human reaction. It is a mechanical reflex. The *human* reaction would be the evolutionary 'what-is-it?' response. Early man looked sharply and intently at his world to perceive what had heretofore been unperceivable. He was concerned with minute variations and impingements that might enlarge his perceived world. His community depended on keen notice of such anomalies.

Being periodically thrown into real time against his will gives the achiever the opportunity (indeed, the necessity) to try something different. For him this moment without a goal in mind could be the crossover point. Were he to live with his anxiety for a while instead of going into a fight-flight syndrome, his life might be radically altered, his mind renewed. But if his achiever alienations seem to pay off moderately well, he will likely continue as before. It would be hard for him to substitute listening-looking for the nervous excitement of a new scenario to exploit a promising possibility. Convinced the visualized tomorrow more real than the unexplored present, he may offer as a rejoinder, "A bird in the hand is worth two in the bush." It just so happens, ironically enough, that the bird in the mind is not the bird in the hand. The bird in the hand is actually the solid unknown that comes knocking in real time and interrupts the familiar image-streams which are in the head, or 'bush.'

The achiever without a fresh goal may feel that his experience is entirely different from the 'disruption of the familiar' which the man of wilderness talks about. After all, he has not wandered about in the past, but completed his schedules in a workmanly manner and arrived at his goals. Everything went as planned and where it did not, he *worked around the difficulties.* His only problem now is determining his next project and who will 'buy' it and set him going again. He feels quite capable of bridging over difficulties; he can handle interruptions and discouragements. But he admits if he cannot locate another goal soon to get him out of his slump, he may need help. With each round of the achiever cycle, his ability to bounce out of today into tomorrow stiffens and loses a little more resiliency. And with growing apprehension, the aging cityman finds himself being pushed ever closer to that 'emptiness' between yesterday and today. For him it symbolizes death, but for the wilderness thinker it has long been his natural habitat.

The man of wilderness is not a man of inaction. But he does not move to be moving. He knows when to pause and look intently and when to move and to search. If a tree is about to fall on him, he is as unlikely as the next to sit and meditate on the eventuality. But where familiarity has broken down, he is ready to hold himself at the point of tension, absorbing the pressure of real time, until that which called notice to itself reveals itself.

Einstein marvelled that the wonder of creation was its intelligibility. Conversely, the marvel of man is his capacity to wonder. He comes with a nature which orients to the faint clues that creation scatters about to tease and tantalize his attention. Thus, there is a correspondence between creation and man's mind. The one is an intelligence whose full magnitude we cannot begin to comprehend. The other is a medium capable of receiving and developing imprints of intelligence heretofore unknown.

Cityculture works to defeat the intrigues between our minds and the unknown. But this mind-to-unknown relationship is endemic to the species. Many a child discovers on his own that he can roll a block across a series of Lincoln logs. In his excitement and involvement he reinvents the wheel. What elements of the unknown the individual will be attracted to and succeed in penetrating, through psychic metamorphosis, become the wonder and glory of his life.

THE UNFAMILIAR AT PLAY

CREATION PLAYS WITH US in the wilderness experience as we might with a kitten. It may start by drawing the merest thread of a curiosity near our path or even across our foot. If we have the wonder to turn and question this least anomaly, our gaze growing fascinated by the slight movements and quivers of the unexplained, the wilderness game is on. When our hand moves to investigate this bit of unknown, another puzzle or several take its place. Each of these leads to even more. As far as creation is concerned, it does not seem to matter where we enter wilderness. Any thread of curiosity will do to get our minds turned in the right direction—that is, towards the not-known. The unravelling of a tiny part of creation then becomes our life.

The wilderness experience begins when the individual accedes to his mind's attraction to the 'unthinkable.' That his reflections will move out past the walls of conventional memory causes him little concern. Rather, to encounter and interact with whatever has no definitions or terms, yet catches his attention continually in one fashion or another, seems most normal. (Whitehead called this sensitive, impressionable mode "the state of imaginative muddled suspense which precedes successful inductive generalization"—that is, seizure by a wholly new perception.)[1] As the individual allows his fascination to lead him, as he plays with it, and as he asks others about it, he will find that their 'answers' do not satisfy him. At the outset everyone seems blind to the wonderful oddness that attracts him. Very often the question central to an individual will begin to surface in its most simple phase at a surprisingly early age. An older Einstein could recall vividly his first enthrallment, at the age of four or five, with the magnetic compass, its needle quivering in response to unseen influences.[2] Later, he traced much of his development of relativity back to that quality of open-ended curiosity which the compass and other intriguingly odd phenomena aroused in him.

Each man represents an entry, never before attempted, into the unknown. His mind, according to this wilderness argument, is preadapted to be one of nature's significant question marks. Each is possessed of some special attraction to the unknown and would be led by curiosity and serendipity into its investigation. In its wilderness mode, this mind is fully capable of relating with the unknown and participating in the psychic metamorphosis process. What continues to catch and tease the individual's mind, like a bramble snagging his sock, will differ from that which catches at the mind of another man. The relative positions of the two will be just different enough to induce special intrigues. As the individual entertains what attracts him (or irritates him) as odd, he is permitting his curiosity to guide him into byways he had not known before. He is no longer under the control of cityculture reality—for his oddity is outside the canons—and he will experience a growth of the mind not known to the achiever. The process begins when we are attracted or upset by what has no ready answer. By responding we carry forward a dialogue with the unknown. And through such intercourse, into which the unfamiliar enticed us, we shall be changed forever.

This self which expresses itself now and then through the intercessions of the wilderness mind is 'non-rational.' Order it will certainly have if the metamorphic process is allowed its course. But the wilderness self will not be a product of cityculture's kit of 'rational' explanations, methods, and values. To be rational, we recall, is by definition to be alienated from the totality of the situation. One cannot be rational—that is, involved with choices—and at the same time engage the real world, which can only be the unfamiliar, the non-codified. The perception and entertainment of the unfamiliar are non-rational, for what has never been familiar or categorized can have little or no utility. If one is to make a breakthrough into an entirely new understanding, the mind cannot be twisting this way and that with thoughts of grading and choice.

The very fact that man's mind can be teased and fascinated by what he does not know immediately suggests that the not-known would seek a relationship with him. Likewise, it reaffirms that our minds are 'by design' capable of response and inquiry into intelligences beyond the bounds of the culture and its familiarity sphere. When man's mind is fixed with fascination in this wilderness mode, he is closer, right then, to his own origins than at any other time. In such brief moments his mind becomes an extension of that larger unknown intelligence that shaped him.

The wilderness mode of the mind, as we noted before, is much like a still hunt—full alertness for what wonder may appear, but total emptiness of choice, utility, expectation. When the communication-perception does arrive from the unknown, it slides in through a back door like a thief. As a dream, hunch, vision, or intrusion of some sort, it may present itself while we are engaged in the most mundane, routine activity. Emerson and Thoreau regularly got their best thoughts during their walks. During a bath Archimedes was struck with the answer to an old problem. Chopping wood, washing dishes, weeding the garden—whatever occupies the hands and body and numbs the senses—have also been occasions when the unknown may slip hints into the mind of inquiry to keep the dialogue going. Daily routines, which are anathema to the dedicated achiever, have often proved the most fertile moments for wilderness adventuring.

Where the achiever mode traffics in isolated fragments—clever cause-to-effect stratagems—the larger, wilderness mind deals in whole images. It has no scenario, no image in mind which it strives to assemble like a craftsman would a predetermined mosaic. Rather, the wilderness mind moves through unfamiliar terrain and stumbles upon that which moments before was entirely unknown. All at once it is presented with a whole image fully realized, much as a curtain is drawn aside and an elaborate tableau revealed.

Ambiguous and broad is this enormity in which we swim and the nature of which we cannot perceive directly or 'on purpose.' It works in vivid images and tableaux whose meanings are uncertain and which could hardly have been invented or anticipated by any individual. The unknown is a game player. Its response may include an occasional word which has been emphasized, but which afterwards may be seen as a witty ambiguity capable of diverse interpretations.

Communications from the unknown, which lies all about us, may at first appear unintelligible, presenting themselves as disturbing riddles. By this movement of life in which there is no familiarity, we can feel quite isolated, even threatened. But the fact remains that the curious thing did attract our attention and perhaps no one else's. And the reactions we had, the hints and insights, did enter our mind and not another's. Every new puzzle has been tailored to our very own discovery potential. To be able to incorporate, even if at rare intervals, some elements of the unfamiliar is to have experienced moments of wholeness.

PHYSIOLOGY OF ACCESS

WHETHER WE KNOW IT CONSCIOUSLY or not, we live in a continuous relationship with the not-known. How could it be otherwise? Sleep studies by Kleitman, Dement, and their colleagues have demonstrated that we dream with great regularity despite our recall being meager. Indeed, Kleitman thought of "the natural state of the brain as resting—the brain must be urged into activity by some agency or need to respond." In his view, it was reported, "the problem was not to explain sleep but to explain wakefulness."[1] Most of us, however, feel quite the contrary: wakefulness, or what we know as achiever thinking, seems clearly enough our natural condition.

Sleep, or the mind/brain in any neutral, unstimulated state, does not seem 'normal,' but a kind of death. When we reflect at all, we wonder at the strange behavior of our mind in that interval of 'non-being' we call sleep. Some of our experiences in this 'dream world' where our choices seem limited or of doubtful effect are amusing. Some, frightening. Some, erotic. Some seem to evolve almost like epics: each tableau translates into another tableau, following an unknown, unwritten grammar. We may even continue the dream 'sequence' later that night or the following night.

We have now learned that in our night's sleep we rouse periodically to a state of dreaming for about 20 minutes and then fall back into a deep, or dreamless condition for another 70 minutes. Clues to the existence of this 90-minute cycle of imaging and non-imaging first came to the attention of Nathanial Kleitman's assistants, Dement and Aserinsky, when they noticed that the eyes of a sleeper under observation fluttered from side to side every so often throughout the night. When awakened during one of these rapid-eye-movement (REM) episodes, the subject recounted a dream he was having.[2]

Afterwards, numerous experiments were made in sleep deprivation and in REM-dream deprivation. In the latter, subjects were awakened throughout the night at the onset of each REM period. Nathanial Kleitman, dean of the sleep-scientists, noted: "Perhaps the most striking unexpected observation was the increase in dreaming during the sleep that followed a period of prolonged wakefulness, a sleep that was thought to be very deep. Sleep deprivation, of course, also means a lack of opportunity to dream, and the occasional dream-like behavior of sleep-deprived individuals could possibly be an effort to 'catch up' on dreaming."[4] William Dement conducted many of these experiments and suggested they indicated

> ...that a certain amount of dreaming each night is a necessity. It is as though a pressure to dream builds up with the accruing dream deficit during successive dream-depri-

> vation nights—a pressure which is first evident in the increasing frequency of attempts to dream and then, during the recovery period, in the marked increase in total dream time and percentage of dream time.[4]

Thus, whether we can recall or not, significant parts of our life are quite accustomed, and fated so it seems, to go beyond the reach of our rational, achiever mind and to play image-making games with the unfathomed.

Related studies in the dream and waking states of man's mind have discovered that the mind has several distinct levels of activity. The charting of the mind-brain is done with the electroencephalograph or 'EEG,' developed in Germany during the 1920s by Hans Berger and perfected by other workers in England and America in the following decades. The lobes of the brain generate electrical pulses which, although minute (on the order of 50 microvolt surges), can be readily detected, amplified, and charted. The latter is conventionally done on moving strips of paper by pens that dart up and down, creating a visible flow of the cycles in that very near, but very distant, world of man's brain.

These waves, interestingly enough, form *four* distinctive patterns as they increase or decrease in frequency. As might be expected, each person produces his own mix and variation of such waves through a 24-hour period. But the four principal waves can be readily distinguished from one person to another. In a way these patterns represent specific activity levels of the mind.

The *alpha* wave (8 to 12 cycles per second) appears when the mind is awake, conscious of the body, but not busy with the external world. It most often occurs when the eyes are closed and the mind is in a state of reflection rather than worry or agitation. The line which the alpha wave makes on the EEG tape is a clean, wide oscillation with only a few variations mixed in. The alpha wave produced by the mind awake, but 'at rest' is often described as the basic wave from which the others are considered variations.

When the individual opens his eyes and his mind becomes active with problems, trying to recognize patterns, doing mental arithmetic, and other strivings after accomplishment, the brain waves speed up and the pattern changes dramatically. The strong and regular alpha waves break down ("desynchronize") into the tight irregular saw-tooth formations which are called *beta* waves (13 to 26 cycles or more per second).[5] One can sense in this change of the zigzags on the EEG tape the abrupt transition from a calm state of the mind to a nervous one.

Below the median alpha wave there appears a wave which, though slower, suggests an amount of activity similar to the beta wave. This is the *theta* wave (5 to 7 cycles per second). Another saw-toothed, irregular wave, it tends to sharp peaks like the beta, but with the slower oscillations of the alpha. In some ways it looks like a beta wave that has been stretched out. Some of the patterns of mixed amplitude in a prolonged sequence of theta waves appear to repeat themselves or nearly so. Few of us are ordinarily aware of this stage of drowsiness or light sleep when the theta waves are produced. We think of this period as transitional: either 'the moment of falling asleep' or the experience of coming awake—both of which seem unimportant, pass-through intervals in our achiever world.

Finally, below the theta and the other varying degrees of 'awakeness' we find the *delta* waves. These appear in full sleep (at about 1/2 to 3 cycles per second), when one is breathing deeply—even noisily—and perhaps turning from time to time. The mind-brain is here in a dreamless condition.

Some of the most engaging work by the researchers has been to determine what the mind is doing in each of these four wave states. After more than 40 years of intense investigation it is now generally agreed that a correlation exists between each wave period and certain characteristic activities of the psyche. Each wave is a kind of signature that indicates what attitude the mind has assumed.

During a 24-hour period man's mind can be discovered operating in any one of four states, with overtones from the others most likely mixed in. Each mode defines, in one way or another, a wholly different category of man's experience as a creature in time. Thus the 'events' which vigorously impressed one in the theta wave dream-state would not likely be the sort that would enter into the mind when it was in the full beta activity of planning, choosing, achieving. The extended, dispassionate, eyes-closed contemplations which might be made in the alpha condition would not be expected in the theta or beta. The selection today of lumber, nails, cement, pipe, electrical wire, and roofing paper to put a house together week after next will most likely done by the mind in beta. But the mind in theta might have had much to do with feeling that the design of the scale model was 'right.'

We seem to have been endowed by nature with two states of animation, or aliveness. In the beta we feel awake and are conscious of time as movement. We know time as past and time as future. In this psychic mode the present is that uncertainty which must be reclaimed—either through adaption to paradigms (re-experiences of sacral time) or by conversion to ideals (apocalyptic time). In this state of 'wakening' we make much use of symbols (words and fractions of the real world) to convert uncertainties to familiarities. And, thereby, we would quantify, divide, and control time.

In the other state of animation, the theta 'wakening'—what Kleitman and Dement called "paradoxical sleep"—real time becomes a player in its own right. The movement from familiar to familiar which we know as 'thinking' is here replaced by movement among unfamiliars. A game is being played and as cityman we do not know how to play. Image dissolves into image. An unexpected scene replaces an unexpected scene. Challenges face us for which we have no 'answers.' Here our usual symbols are of no help to achieve our way out of the impasses and embarrassments. On this stage whole images are interacting. Time seems collapsed: we are experiencing, but without past or future. To be sure, we seem to breathe, to move about, to feel, and to exist. But all in a time frame unfamiliar to us. Since our culture has centered our minds in the alienated worlds of time past and time future, this real-time world appears bizarre to us. On awakening, we dismiss it readily, even if we cannot always find it easy to forget.

IN THE SIXTIES A NEW CLASS of researchers—important among them, Barbara Brown, Joe Kamiya, Elmer and Alyce Green—discovered that Western Man, with a little biofeedback training, could with his mind perform certain exercises that the Eastern yogi and the American Indian had practiced for millennia. He could remove his attention from the active, or beta, level and for varying periods place it in one of the other waking modes—alpha or even the slower theta.

Brown in her *New Mind, New Body* and the Greens in their *Beyond Bio-Feedback* described with engaging anecdotes how feelings and thought experiences modify and create signatures in the brain waves. When the alpha waves show on the tape, the mind is probably in a reverie state, reflecting on a subject rather than deciding on a course of action. Likewise, if one relaxes after some difficult work, further decisions put aside, the mind will likely slow and the alpha waves will begin appearing—especially if the eyes are closed. Brown observed that if "an individual focuses his eyes for precise visual attention, alpha disappears, and conversely when the eyes are de-focused, alpha reappears."[6] She adds, "Almost any sensory stimulus, whether visual, auditory, or tactile, expunges alpha from brain electrical activity, particularly if the stimulus is intense or sudden."[7]

Slower than the alpha, the theta wave is nevertheless an active, much modulated signal. When this form appears, the individual is most likely dreaming or experiencing some similar encounter with images. During the rapid-eye movement periods, the theta wave shows up in abundance. The energy that is expended by the brain in this mode seems remarkable when we remember so little of what went on—and when we presently have no conception in our inherited culture of the importance of this function. As far as we are concerned the theta waves and the dreams that go with them might as well be an appendix, one of nature's anachronisms.

Each of the several modes of the mind was found to have its own specific 'feel.' And within each feeling certain thought experiences were more likely than others. Often it was enough, the experimenters found, for the subject to concentrate on the sense of a mode in order to place his mind-brain into that level. As the brain-wave output began registering the desired wave patterns, he would begin experiencing the thoughts characteristic of that wave. Thus, mind-imaging, brain waves, and body response-activity are all three intimately tied into each other. And these were more varied within their modes than we achievers had commonly realized.

Our concern here, as throughout this inquiry, is with the relationship between time and mind. If we look closely, it becomes evident that these several aspects of the mind and the corresponding wave forms are further depictions of the time-mind equation. When the mind exhibits the beta wave, it is engaged in the use of known symbols. With these symbols it would master time through alienation—creating either a past revisited or a future to be experienced. Clock time and calendar time become very important. Resources must be found and allocated. In beta the mind fashions its own world to step into—separate from creation and its time.

The mind in alpha, however, is man quietly thoughtful, experiencing, but not sensing, content with reflecting on whatever curiosity has intrigued and holds it. Time as a distance being travelled, as a resource being expended, is sharply reduced in significance.[8] To the mind, when in a condition of wondering and inquiry, an hour can seem a minute. Here, time has lost its linear appearance and become a global absorption in the wonder and realities beyond the scope of words. The mind is centered not by an achiever goal, but by the charm of the unfamiliar. Fascination alone turns down the sensory inputs. In alpha the psyche rests on the threshold of the unfamiliar.

The mind moving into theta crosses that frontier. It gives up on the language of symbols and, going directly to the project at hand, plays with the unfamiliar. The mind, we believe, is preadapted to encounter and to grapple with fields of intelligence beyond its everyday reach. But it can only perform this interaction with the mysterious and unperceived in real time. And this means that the usual alienation of time by the senses and by the wordplay which some call "directed thinking" must give way.

Words are based on the familiar—recognition and use. But the present business is to compass the unfamiliar. We have long attached our sense of individuality to this mind of reason and choice, of words and values, but now discover it severely flawed. The citymind of the achiever is simply incapable of sensing or relating to unmapped, undiscovered fields of intelligence. The dimensions of such an expanded reality can only be felt out in a spontaneous play of whole images. And these must be experienced by the mind at a time-level other than the familiar and measurable.

The natural mind will not be diverted from its primary function any more than our lungs will permit us to stop their compulsion for a gasp of air. The mind is the field on which it seems we are destined by nature to play with the unfamiliar. To refuse the game, to duck and run, is a disavowal of mystery—that is, those extensions of creation which lie just beyond our usual senses. Each rejection of the invitation to wonder is a victory for cityculture and a further wounding and scarring of the mind-brain.

Completely apart from our 'will' to order the familiar, however, the natural mind is busy with the not-known. Since our words extend only to the recognizable and doable, the natural mind must use images and displays. Most of us have restricted our lives to the familiar and to the waking moments when words reign supreme.

As a consequence, our mind's ongoing engagement with the unfamiliar appears with any regularity only in the periods of paradoxical sleep. Yet, as most of us know, these image-plays can, if they so choose, impress themselves on our 'waking' moments with great vividness. And such occasions can occur unexpectedly whenever the unfamiliar collides with our schedules and when the orienting (wilderness) response seizes our minds.

THESE WAVE-ACTION STUDIES OF THE MIND have been broadened by several related investigations. The first describes the activity of the dream-theta mind during our waking day. Another links deep meditation to the theta wave. And the third considers what happens to the wave movement when the mind is confronted with the unexpected and responds with the orienting reflex. All three are of considerable interest to our inquiry into mind as time.

The initial study indicates that the curious REM sleep cycle, first described extensively by the Kleitman laboratory, may not end with the break of day. There is evidence that the 90 minute period carries right on over into the waking condition. During the day it becomes what has been termed the basic rest-activity cycle, or "BRAC."[12] And within this essential hour and a half period there appears a natural, involuntary surge of activity very much like the REM period at night. In cityman such daytime intervals may evoke restlessness—a desire to eat and to move about, and in general, we might say, a *dissatisfaction with the familiar*. If we are aware of this 'nervous,' unplanned activity—image-play at night and disquiet by day—we may attempt to fit its expressions into some acceptable cultural context. To our dreams we would attach an appropriate 'meaning.' During the day we would calm our 'nerves' with a cigarette or a cup of coffee. Some of us may at odd moments draw cartoons; others, write lines of funny doggerel. What we appear to have stumbled upon, in these important studies of our psyche by night and by day, is the threshold of the autonomous mind—that mind which would be 'busy' with the unfamiliar.

Heretofore, this cycle of increased energy, as described in the BRAC concept, has been largely undetected—much as the moon in daytime is made invisible by the brilliance of the sun. This recurring restlessness may be translated as a periodic return of the mind to its natural state, interfacing with the unfamiliar. It would have relief from the routine, the ordinary, the encompassed, the fictitious. The impulse to move about, the desire for a short nap, the blinking of the eyes, moments of reverie (as when driving or doing chores) might well be considered as additional 'intrusions' of nonlinear, nonverbal experiencing. That which is routine in sleep, the lapse into theta-imagery, may also be ready to express itself with micro-moments of subliminal imagery during the day. While the mind is busy with linear time, as tomorrow (beta-achiever), or as yesterday (beta-traditionalist/cultivator), the unknown would periodically pull the mind back into a time without movement—time as contemplating and observing, and time as enriched with undirected image play. As the blink of the eyelid periodically restores moisture to the eye, the fall-back-into-the-dream-reverie would restore the mind-brain to the reality of a non-alienated world.

The field of the unknown is like a great forest at night. We collide with unseen masses and trip over what we do not know is there. Without our conscious knowledge or bidding, our mind is ever at work on these encounters, wondering, watching, inquiring, and flashing its responses. For most of us, who know only the beta mind, such work goes on 'out of sight.' It is carried on by psychic activities able to work with unfamiliar images in a time configuration of which we have been made 'unconscious.' To understand that mode and to experience its rhythms is our purpose.

What happens in the mind-brains of individuals absorbed in meditation has long been a wonder. Studies were conducted by Elmer and Alyce Green at the Menninger Clinic. For many it seems the meditator is wasting time when he or she could be active and doing something 'creative.' The Greens found expert subjects who would allow their brains waves to be measured. These EEG testings of adept meditators—Zen monks and Indian yogis—have shown repeatedly that the first level of meditation produces alpha waves. But when the point is reached where the senses are turned off—yet an inner consciousness is still awake—the theta wave is quite likely to occur.[10] The mind has passed from the world of symbols to the world of whole images, from alienations of time to the realm of real time.

In the past, relating intimately with the unfamiliar and experiencing corresponding expansions of consciousness have been described as exceptional religious experiences rather than as normal functions of man's mind. The purpose of the ancients—whose tradition the meditators carry on—was to access that mind which as cityman we have all but buried. For it is this mind that is equipped by nature to relate with the yet-to-be discovered.

The third fascinating discovery which the investigators and their EEG machines have come up with is the connection between the orienting reflex and the theta wave. Ordinarily one would expect that when one experienced the onset of an altered situation—the sudden noise, the flash of light, the loss of keys, the collapse of any expectation—that the mind would respond with a burst of beta activity. And this does appear to be the initial response in the "what-is-it?" reflex.

When the interruption does not go away, however, but is repeated, the mind tends to move in the *opposite* direction. Investigators call this secondary response the "inhibition" of the original, heightened, orienting reflex. The brain waves slow down and enter the theta region. A kind of 'drowsiness' occurs and the initial reaction to the stimulus is no longer observed. And that may be the moment which we will later describe as, "I felt a little faint at that point and had to sit down." (It must also be noted that the orienting response/reflex is a whole body reaction: the heart, the lungs, the skin, the proprioceptors, as well as the brain, are all involved.)

The Russians since Pavlov have been keenly interested in psychophysiological studies. A. N. Sokolov, one of the leading orienting-reflex investigators has observed: "The development of inhibition of the orienting reflex is accompanied by changes in electrical activity from the beta-rhythm to the alpha-rhythm and so on towards the even slower theta-and delta-rhythms."[11]

And what is going on when the mind 'retreats' to the theta level? One might suppose that this inhibition was due entirely to the subject's 'turning off' the intrusive signal. But if that were so, would the brain waves necessarily sink towards the theta level? Might they not continue in beta on the subjects the mind was originally engaged with? One could also wonder if the inability of the subject to handle the repeated intrusion did not occasion an unconscious retreat of the mind to a position where it was better equipped to consider the undefined and unexplained. Barbara Brown, an American researcher, made this valuable observation: "Theta waves have been found to be associated with several different behavioral functions such as drowsiness, dreaming and orienting (particularly the assimilation of new information)."[12]

In this brain-wave research, in our anthropological studies of early man, and in our review of the uncommon-man phenomena, we are finding that evolution may well have created us for a different purpose than we heretofore suspected.

Our initiation into the city world as an adult consists of becoming goal-oriented. We learn to isolate resources, to make choices, to build scenarios, and to push them through to completion despite all obstacles. This is the essence, we have learned, of the "can-do" hero.

All the while, however, it is not to the familiar that we respond, but to mystery—that which lies beyond the "can-do." Not only our mind-brain but our whole body reacts and comes alive in the presence of the unfamiliar. We call out, "What goes there?" Our eyes dilate, our heart beats faster, the hair on our skin prickles. After the initial burst of energy, when the fight-or-flight mechanism is in control, the mind slows into its mode of search and listen. It is here that the mind would reach and encompass that for which it has no ready images, words, or preconceptions.

Thus far, then, our researchers have shown us that our total mind is more complex and suggestive than our cityculture apologists have allowed for in their utilitarian philosophies of mind as linear tool. Important areas in our mind are quite comfortable with time as present and non-achiever. In fact, they do not seem to 'know' tomorrow. The achiever surface of the mind is surprisingly thin, fragile, and artificial. The larger, resilient faculties of the mind lie just below, in the regions of alpha and theta energy play. The mind is clearly preadapted to consider the unknown, to be attracted to the little strangenesses by which the unknown would open a dialogue. And the 'restless' cycles, night or day, seem to be significant channels through which the unknown may relay its image responses. The electrifying 'break-in' of the vision experience, whether as vivid dream-fascinations in our sleep or as micro-flashes of insight in our waking moments, is not aberrant, but somehow an essential part of the mind activity.

IF A STRONG CULTURAL AVERSION towards the unfamiliar were implanted early in the child, the individual might never incorporate what the alpha-theta levels of the mind seem intended to supply—namely, his life-enriching experiences with reality systems larger than his present level of perception and interaction. Indeed, he might live his whole life unaware of the other activities of his mind and what they would do for him.

Retardation can be brought about in man by a variety of methods. His brain can be damaged at birth. Or, later in life, he may experience a severe concussion. He may breathe noxious gases or ingest addictive chemicals. All these can have a degenerative effect on the brain and thereby diminish the range and quality of his faculties.

The mind-brain is, we know, in an especially sensitive condition when we are young. It is then that many of our thought processes and attitudes towards the environment we are fated to live in are formed and set. Shock experiences and powerful aversions create psychic scars. Early on, we become self-censoring: to get along, we go along. Every dominant historical culture, cityculture among them, has had its cabinet of horrors. If the individual ventures beyond the prescribed norms, fears will paralyze his mind and heart. Imprinted in the marrow and brain-cells of every cityman is the aversion to quitting cityculture's orientation to the familiar tomorrow.

That an entire generation of psychologists would become involved with 'behavior therapy' and 'behavior modification' was not accidental. The discipline of social engineering includes operant conditioning, aversion therapy, and desensitization. We know control by manipulating the familiar, in its more overt form, as advertising and propaganda. But cultural management of man's behavior is ancient. It is built into the definition of time which runs like a common nerve through everything in a given society. Today, the methods of instilling and enforcing the code of what shall be familiar have been considerably refined.[13] But there is remarkably little that is subtle about the purpose and reach of these highly educated, highly motivated enforcers.

Professor James V. McConnell was head of the Department of Mental Health Research at the University of Michigan. In his odd moments he edited for the experimental psychology field a humorous newsletter, which he called *The Worm Runner's Digest.* McConnell's perhaps most well-known thesis concerned the transfer of intelligence through digestion. He could show you startling evidence that an untrained worm (planarian), if fed the carcass of a trained worm, would 'inherit' thereby some of the intelligence of his repast. But consider here, if you will, a sample of the lines

Professor McConnell wrote elsewhere about modification of 'social deviants':

> It is axiomatic in the behavioral sciences that the more you control an organism's environment, the more you can control its behavior. It goes without saying that the only way you can gain complete control over a person's behavior is to gain complete control over his environment. The sensory-deprivation experiments suggest that we should be able to do exactly that.
>
> I believe that the day has come when we can combine sensory deprivation with the use of drugs, hypnosis, and the astute manipulation of reward and punishment to gain almost absolute control over an individual's behavior. It should be possible then to achieve a very rapid and highly effective type of positive brainwashing that would allow us to make dramatic changes in a person's behavior and personality....
>
> We should reshape our society drastically so that all of us will be trained from birth to want to do what society wants us to do.... Today's behavioral psychologists are the architects and engineers who are shaping the Brave New World of Tomorrow.[14]

Early aversions followed by periodic reinforcements result in notable losses in the functions of the whole mind-brain. In cityculture the major target for repression is the mind's natural play with the not-known (the anomalies of the moment) which begins very early. Again and again, the time-harassed parent or older sibling says, "Don't touch, don't touch!" And later: "You're asking too many questions!" The process continues in the academic channelling where one must become familiar with what others discovered long before.

Retardation and restriction to the well-known—through induced appetite for tomorrow and aversion to today—are almost without exception successful. (Remember McConnell's words: control of environment leads to control of behavior.) Such desensitization is nearly indistinguishable from a kind of simple lobotomy. The individual gets a job, cares for himself and seems operational. But certain activities and areas of perception that ought to be normal—such as play with the unfamiliar—are effectively blocked. Physical damage to the brain of a radical degree is irreparable. Retardation through cultural damage is nearly so. Aversions to the unfamiliar are not easily lifted when they are reinforced by every cultural import minute to minute, year by year.

The mind of cityman is so affected by cityculture and its tomorrow orientation that it 'lives' in two states only, the hyperactive beta and the insensible delta. As a living entity, cityman conceives of himself as fully awake and running in his hot-blooded achiever pursuit of tomorrow. Otherwise, he is fully asleep, dead until tomorrow presents itself as today again. When he awakes, he may say, "I slept like a log!" The existence of two other levels of the mind (the alpha and theta) between full stop and full acceleration—with their own special activities and inputs—is far from obvious to cityman. Earlier in our century, we find even so acute an observer as Jung making his observations of these "in-between" minds. He dismissed them as "archaic."

> Archaic thinking is a peculiarity of children and primitives. We now know that this same thinking also occupies a large place in modern man and appears as soon as directed thinking ceases. Any lessening of interest, or the slightest fatigue, is enough to put an end to the delicate psychological adaptation to reality which is expressed through directed thinking, and to replace it by fantasies. We wander from the subject and let our thoughts go their own way.[15]

The wideawake achiever and his "directed thinking" is man's adaptation to a tomorrow-reality. For most, then, the in-between states of the mind-brain, the alpha and theta modes, simply do not exist. These, together with low-grade achiever musings about tomorrow's pleasures, have all been lumped by the early Jung into the pejorative category of "fantasies." The reason for blacking out alpha-theta by means of aversions is that these in-between borderlands of the 'unconscious' involve aspects of time which are not citytime. For it is here in alpha-theta that time becomes play and perception—challenge and watch—rather than gross movement and exploitation. When cityman has reveries, vagrant hunches, and moments of wondering at the curious, he makes haste to dismiss them as a "waste of time" and tries to get on with the "real world"—i.e., scenario planning for his tomorrow and for his days after tomorrow.

Earlier we described the "orienting reflex" which corresponds in ways to our own model of the wilderness instinct. When the unusual occurs to man, he responds with inquiry. He seeks to discover "what is it?" In the orienting reflex it was noted that if the intrusion continues to "present itself," the reflex proceeds into a secondary phase known as the *inhibition* of the orienting reflex. Responses diminish and the subject grows "drowsy." The brain wave slows into the alpha and down into the theta region. In our wilderness model we have suggested that the mind, when confronted by the unknown, commences a cycle the second phase of which is an abandonment of the familiar and an immersion in the not-known of real time. Such would be the response of the healthy, undamaged mind-brain.

But further studies in the matter of the orienting reflex shed light on what can happen when there is damage to the brain. Where the cortex has been removed or chemically anesthetized in laboratory birds and animals, the orienting reflex is affected. The first stage, the initial response, goes off as expected. The creature notices and comes to attention. But in the second stage, when the intrusion (the sound or light) is repeated, the inhibition and "drowsiness" do not occur. Every ringing of the bell or flash of the light is followed by the first stage. There is no diminution of the "coming to attention." In humans this same curtailment of the reflex has been noticed among birth-damaged infants or those with defective brains.

> Apart from normal children, the children traumatized during birth were investigated. In this group of children the reactions to the stimulation of the analyzers were produced in the same manner. Some of the children in this group sucked badly and reacted feebly to the stimuli, but a greater part of them reacted sufficiently distinctly. In many of them the reaction did not show a tendency to extinction even after 10 presentations of the stimulus....
>
> As can be seen the children who have experienced a birth trauma show a definite deceleration of the speed of extinction of the orienting reactions. This is a good place to cite the ... observations by Itina on prematurely born babies in which it was frequently not possible to achieve extinction with repeated presentation of sound stimuli....
>
> These results can be related to certain observations by Pavlov on animals in which the cortex of the hemispheres was removed through an operation. In these, as is known, there occurred no extinction of the reactions to the stimulation.[16]

What does this mean? It would appear that certain traumas experienced by the brain can prevent the orienting reflex (the wilderness response) from passing into its second stage. The brain so damaged is no longer capable, it appears, of slowing down into the theta range—and possibly accessing that mind which investigates the strange and undefined. If all of this is so, then one wonders what has happened inside the heads of those adults who keep on responding to alarms with no "inhibition" of the reflex? What are we to make, for example, of the maximum leader who makes a career of blowing his trumpet and calling for increased militarism? And how is it that each time he does so, the masses jerk and twitch in unison and vote for more planet-killers as though they were hot-wired. The ability of Hitler to affect masses of people was the extreme example. But since then many similar 'leaders' have used these techniques with greater subtlety—the same totalitarian purposes of managing large populations being served. Is it not important to wonder how our mind-brains have been so desensitized that they are incapable of slowing into the more reflective modes? Where, then, are those undamaged minds still capable of reevaluating such repeated hysterias—and of redefining the world and the meaning of man anew? Without this wilderness depth to our thinking we are little more than markers, or pawns, to be pushed about and manipulated by resource-using powers that would define time for us.

Man commenced as a hunter-gatherer. He evolved over several million years that we know of as a crafty, resourceful creature, able to deal with the unexpected. His mind was capable of functioning, at some time in a day-night cycle, on four different levels. Each of these levels, or modes, gave him certain resources in his relationship with his environmental setting. About 10,000 years ago the alienating forces, inherent in the 'newer,' planting culture, began to take over from the natural evolutionary processes. They have been steadily at work ever since, constructing their airy realities outside of the real-time world.

By now these powerful formatting traditions have created their own version of man. The mind he is born with is still the four-level mind of the original hunter. But by the time he has reached adulthood and is recognized as a cityman, his mind will have been partially disabled. The whole mid-range reflective apparatus, which enabled original man to become the splendid creature that we know he can be, is in effect stripped out, like the entrails of a fish. Cityman is a pathetic cripple who struggles as he can, alternating between the two levels left to him—awake in alien time as achiever or dead in a dreamless sleep. He has lost that which essentially makes him human, for he knows nothing now of real time and his special capacity to think in a non-alienated mode. Fortunately, the truth is that his damaged mind is not really lost. Only paralyzed with fear.

Everywhere the extraordinary is happening to us, yet we continually translate it in terms of the ordinary. But if the mind should begin to find delight again with small, unthreatening oddities that tease one's attention and may even sidetrack one's plans, some quite wonderful things can happen.

THRESHOLD OF THE UNKNOWN

MAN'S DREAMS ARE GREATLY VARIED as everyone knows. We have dreams of danger, of seduction, of being lost, of being unable to meet expectations. Whole industries have come into existence, early and late, purporting to make 'sense' out of our dreams. No matter whether they are rambling anxiety dreams or the gem-like vision, they are subjected to an analytic process that seeks to reduce them to familiarities. Ancient soothsayers listened to the king's premonitions and dreams and advised this destiny-changer what lay in the future. And, in the absence of a royal nightmare, these worthies were ready to peddle their own 'visions' to the court.

Our psychiatrists have made it a vogue to find one's childhood, or yesterday, replayed in one's dreams. We are learning, however, that the dream experience may also be influenced strongly by the present. Each doctor has his own expectation of what comprises a significant dream. His anticipations are transmitted, subtly or with considerable suggestion, to his clients. The mind, when strongly impressed with expectations of someone it would please, is thrown into the achiever mode. The dream becomes contaminated and may respond in kind. Thus, on the Freudian couch, you were likely to have Freudian oedipal dreams. For Jung you would have encounters with the Jungian archetypes. And with Adler your will to dominance, life-style, and organ anomalies became the stuff of nightly uncertainties. Jung told two separate anecdotes which give us some insight into how compelling this hypnotic effect can be, even on the psychoanalyst himself. Jung was highly amused, on one occasion, when he made a strong statement to a colleague and then noted how that hapless man parroted Jung's opinion even though it was contradictory to his own well-known beliefs. In another situation, however, Jung came under similar pressure to fulfill powerfully expressed expectations. This time Freud had set the determinations and Jung had to extricate himself. Jung confesses to us that he resorted to a hasty 'invention' in order to explain away one of his own dreams which might have incurred Freud's displeasure.[1]

Since those decades when Freud, Jung, and company laid claim to the study of dreams by right of discovery, we have learned something of the dream investigations which early people regularly engaged in. Our psychoanalysts, it appears, are latecomers. The Senoi, a 'pre-literate' tribe of the Malaysian hill country, were studied in the 1930s by Kilton Stewart, who marvelled at the sophistication of their dream studies. These 'primitives' did not discount, as westerners are wont, their nocturnal psychic adventures: "Breakfast in the Senoi house is like a dream clinic, with the father and older brothers listening to and analyzing the dreams of all the children."[2] Rather than be an object in the dream to which things happened as in our experience, they were encouraged to interact. If a terror pursued them, they at first might flee—and wake up. But on a later occasion—after much encouragement from the elders—they would turn while in the dream and challenge that which frightened them. The hostile images represented personal conflicts, they believed, which the individual has not addressed. Here, we might speculate: the disturbing images, while having personal antecedents, might also represent collisions with larger, unseen fields of intelligence that were calling attention to themselves. Kilton Stewart again:

> My data on the dream life of the various Senoi age groups would indicate that dreaming can and does become the deepest type of creative thought. Observing the lives of the Senoi it occurred to me that modern civilization may be sick because people have sloughed off, or failed to develop, half their power to think. Perhaps the most important half. Certainly, the Senoi suffer little by intellectual comparison with ourselves. They have equal

power for logical thinking while awake, considering their environmental data, whereas our capacity to solve problems in dreams is inferior compared to theirs....

In the West the thinking we do while asleep usually remains on a muddled, childish, or psychotic level because we do not respond to dreams as socially important and include dreaming in the educative process. This social neglect of the side of man's reflective thinking, when the creative process is most free, seems poor education.[3]

If life has indeed preadapted us for relationships with the uncomprehended—with fields of intelligence beyond the culture—perhaps we should not be too surprised at the interruptions we experience of our expectations and prearrangements. These could be taken as possible signals that such an unsensed field had been encountered. Normally, when the unknown strikes, we back off and avoid its difficulties and disruptions. We increase our commitment to the familiar in any of a hundred ways (have a snack, talk with our gossip on the telephone, read a magazine, turn on the TV, go shopping). If our curiosity, however, were aroused sufficiently—to the same degree that suspicious noises at night will not let us sleep—we would hold ourselves on full alert for the source of the disruption. This sharply focused inquiry relationship with the unfamiliar is the non-achiever mind at work. And, it happens in another area of the psyche, as we have learned, than the achiever-beta and its usual strings of familiar symbols and metaphors. When awakened by the sudden noise, we might think for a split instant of dashing out in our pajamas to attack the intruder. But then we think of the potential danger. At that moment our mind has no tomorrow to jump into. The so-called flow of time has jelled in the chill of the night. And the processes of our thought now 'fall into' a watching, expectant, timeless mode to discern what is being 'overlooked.' If, as we have said, the disruption proves simple—the noise at midnight being a branch hitting the house—then the achiever level will quickly take over with a little scenario to fix the problem. And things will be back to normal.

Some interruptions of the familiar, however, are more difficult. One may spend years trying to sense the contours of a really profound anomaly. If the achiever's mind has some flexibility left, he may be drawn into a series of interesting steps. The progression begins, of course, when the day-to-day routine is interrupted. The paradox, the unexplainable, intrudes. The mind busily tries this explanation and that scenario, but all usual patterns and resorts come to naught. The stream of familiar images we know as thought slows. Watchfulness and study take over and attempt to locate the 'mouse in the machine.' Time no longer passes, but creeps along, awaiting the discovery of the 'glitch.'

When the unseen field of intelligence still refuses to yield—demonstrating itself beyond handy symbols, conceptualizations, grammar, and comparisons—the ordinary, busy mind loses its capacity and becomes stymied. Now both achieverism and examination (recognition) have failed. Time has not just slowed, but becomes transformed into another kind of time which does not move at all. Upon the failure of the conscious, decision-making level of the mind, other modes are alerted with which we are poorly acquainted. The uncomprehended having shown itself intractable, all psychic energies and powers respond to this challenge. And the individual, feeling a loss of everyday consciousness, may feel faint, lie down for a nap, or wander into the woods while his nonlinear, non-rational, imaging mind takes over.

The profound paradox works to center the mind in a time between yesterday and tomorrow. For an individual accustomed to words and plans, those bricks and mortar of the future, the result can be a dreadful shock. Though in a waking state, he feels himself in a dream-nightmare, teetering on the edge of a cliff. Can he fly or will he fall? His mind must somehow 'work,' but without using words and preconceptions. It can only resort now to images, games, tableaux, and other direct encounters with the unfamiliar.

We might say, therefore, that to dream is to encounter and to encounter truly is to dream.

In such a dream-encounter state, the mind is freed from the tangle of words and grammar used in the time-as-movement world. In the latter realm, the mind has been concerned with those 'realities' that words enable man to create. In the dream experience, however, the mind divorced from words (and tomorrow) can probe into that which is larger than words and is everywhere true. Images and situations that may puzzle the tomorrow-oriented mind, the 'dream-encounter' mind uses with confidence and even genius. In this 'hidden' part of ourselves we are already accustomed (pre-adapted by nature) to engage and relate with the not-known.

The absorption of the individual by intense wonder—through his collision with a large, unmapped field of intelligence—parallels in some ways one's falling blindly in love. The total challenge rallies the protean, inventive mind which likes nothing more than to game-play with the not-known. Once centered on a paradox, this mind puts out feelers, so to speak, to estimate the dimensions and nature of the unfamiliar. Of itself the dream-encounter-discovery mind is not concerned with making alterations. For alterations are the taint of the achiever level and its known symbols for tomorrow. The dream-reverie level works beyond the known, outside the limitation of words and scenarios, in the fine passage between past and future. It would search out the timeless precepts and 'laws' of that which confronts us, just as Newton's mind was aroused to find the ever-present meaning of gravity. In short, the nonlinear, non-temporal mind, awakened by mystery, begins of its own volition to play with the unfamiliar as only it can.

For most of us, whose minds have been locked into the directed, achiever mode and into the notion of time as the better tomorrow, our dream experiences are forgotten before we awaken. They are the weeds of our night mind. And our reveries are dismissed as woolgathering and "archaic thinking." Yet, much of the work done by our most powerful mind, on problems reaching to our very core, is reflected now and then in our dreams, near-dreams, reveries, and hunches. By ignoring this image play with the unfamiliar, we not only repeat old problems endlessly, but also starve the mind. It is just such engagements with the not-known which provide the psyche with its basic nourishment.

Encounters with the unfamiliar—the achiever disabled and the 'dream' play mind triggered—can become a special time of productivity far more intense and rich than any achiever interval. We know, for example, that the blind John Milton composed much of his great epics, *Paradise Lost, Paradise Regained*, and *Sampson Agonistes* at night while asleep. In the morning he would wait anxiously until his daughter came in to take his dictation of the 30 or more new lines and unburden his mind. He felt like a cow, he remarked, that waited to be milked.

Robert Louis Stevenson, while a college student in Edinburgh, began "to dream in sequence and thus to lead a double life—one of the day, one of the night—one that he had every reason to believe was the true one, another that he had no means of proving to be false."[4] So familiar did he become with this inner stage on which the unusual, surprising, and thrilling were enacted that he would challenge this inventive-protean level of his mind with a problem before sleeping and be fairly certain of some unforeseen answer by morning. He referred to this inner mind as his "Little People" and his "Brownies" and confessed they "do one-half my work for me while I am fast asleep, and in all human likelihood, do the rest for me as well, when I am wide awake and fondly suppose I do it for myself." This latter seems clearly enough to refer to intrusions of insight, those spontaneous inventions and epiphanies that seem to come out of the blue and 'hit' us in the midst of our other activities.

The sleep scientists have suggested that the REM experience may have its daytime analogue. The irruption of image-play would then appear to be fairly continuous, more obvious by moonlight than in the glitter of the full sun. Our inner mind that uses images and situational drama to interact with the not-known does not really start and stop. It is ever at work on the unfamiliar. Both by night and by day it is trying to feed us hints and pictures of that with which we are confronted.

The relations we as cityman have with our 'dream mind' seem few, mysterious, and ineffectual. This faculty which Coleridge was wont to call "imagination," others calling it "fancy," we are referring to as the "wilderness," or discovery mind. The natural capacity for hunches, visions, and impressions, which would open us to the not-known, appears to us much like an aunt we saw a few times as a child. The family describes her as dotty and makes fun of her postcards which she sends once or twice a year.

Cityman may claim he has no dreams and that he has corralled his irrationalities. Yet, we may observe daily this same man projecting into his environment all that dream-image play he will not acknowledge.[5] What the adult man of wilderness resolves as an interior experience, the 'adult' of the city turns outside. He is the man who quarrels with his lover at the Laundromat and strikes her about the head. He is the man of *machismo* who sees security in weapons and force rather than arising from intuitions which alter situations altogether. He is the maximum leader who looks upon every peasant group and local cooperative as the masses conspiring instead of an ingenious and worthy people improvising. The achiever 'spills his guts,' as it were, and exhibits for the whole world to see his dangerous, retarded mind. Once the limits of his tomorrow-fantasies have been reached, cityman tends to exhibit an angry-baby violence, towards himself and others.

Nature intends, it would seem, that we encounter the not-known within ourselves and expand our perceptions thereby to mature as the shifty, adaptable creature we really are. Ordinarily, fascination and curiosity would call even the achiever's attention to this internal image play. Dismissed and denigrated, however, man's subjective image-experiencing is set free to interdict in any manner it can. The dream-encounter mind, denied its normal discovery activity, may join the senses in their game of inversion and projection. Just as the senses may cast onto the environment what the achiever wants to experience, the dream-mind, as interface with the not-known, is now at liberty to throw all manner of hallucinations before cityman's eyes. The images he should be observing and inquiring into internally now set upon him externally.

The archetypal cityman would involve the community in his personal resistance to the not-known, even using the planet as his battlefield. He lives with a severely diminished mind and expects everyone else to do the same. Rather than facing his inner nighttime tigers, this cultural clone sees them in his waking life charging in at him from all directions. If he had allowed his perceptive-inventive self its way with the unfamiliar, these tigers might have dissolved. But now he has mined the globe with nuclear poisons and explosives for his terrible tomorrow.

WE HAVE DESCRIBED THE MIND as a function of time. The hunter, for example, perceived one kind of time and all his thought was so affected. For each of the several time-modes in which man's mind has functioned there seems to be a corresponding 'dream mind,' or at least a series of characteristic autonomous image activities. Thus, the hunter who lived in time-present was quite aware of the importance and value of intimations and hunches. The activity arising from contact with the not-known, he called his 'dream-power.' He thought of his dream-discovery mind as his tutelary companion, as his guardian, as his grandfather. Likewise, the spectacular herds, flights of birds, and runs of fish with which he shared life were seen as reservoirs of mystery-power, archetypes of the unfathomed.

The cultivator who displaced the hunter lived consciously in his paradigms of the past. Whatever happened today was suspect until it fell into line with some event in the sacral yesterday. He must not listen to his own hunches lest he contaminate himself.

Whenever he heard tales of the hunter's dream mind and its strange powers of intuition, he became terrified. He believed implicitly that the hunter-shaman knew the secret of flying and could, if desired, steal his soul. Confrontations with the not-known, the peasant dreamed of as encounters with sorcerers, snakes, and archetypal gods. The priest was kept nearby to exorcise outbreaks of the unfamiliar.

The cityman, as we know, simply dismisses the dream-mind. Freud, Jung, Nietzsche, and others assured us that the mind which deals in images of the unfamiliar is merely a remnant of the original chaos. The achiever has learned to desensitize and displace the dream mind with circuses and violence. Cityculture is so contrived as a mazeway of permissions and aversions that no right-thinking cityman can see himself in a direct, ongoing encounter with the not-known. His perception of environment as familiar and usable is the basis of his 'intelligence.' His reality is to give this known field a new and 'better' order. What is not-known might be an interesting oddity. But the reason for its being unknown, he will argue, is most likely because of its low worth as a resource. Any attention given it, therefore, could only be a waste of time and diversionary. More important are the inventions of new uses for the already known.

In consequence of this mind-set the natural relationship with the not-known is not only avoided by the achiever, but 'left to its own devices.' There are occasions, however, when it would appear that the dream-image-mind comes to the aid of the achiever. He is always on the outlook for new scenarios that will give him an edge over others in assembling and converting resources. And when a string of ideas floats in from his memory—which is not to be confused with the dream-image-mind—he feels himself creative and 'in touch with his subconscious.' Is it little wonder, then, that when the real depth mind slams him up against a shocking image of what he is avoiding, he is frightened for his sanity? Such an encounter with the not-known might be a dream of opening a coffin, discovering a rotting corpse and finding it to be himself.

WHILE OUR DREAM-IMAGE PLAY is generally forgotten on waking, a few night dramas will not be dismissed or discounted. So special, vivid, and pointed are these happenings that they can fetch us, all at once, wide awake. Panting and perspiring, marveling or scared witless, we lie in bed a long while, wondering whether we dreamed or whether the events in fact happened. Some of these exceptional dream-visions are recorded for us throughout the Bible. They are clothed in the language of a religious, exceptional occurrence. And thereby their reality as a normal and necessary function of the human psyche, even if unexplained and mysterious, is obscured further. Some of these wonderful happenings we shall discuss in the forthcoming essay entitled *Life Of The Mind.* There we gather many examples of the wilderness mind at work. Despite our being cityachievers—holding ourselves superior to our 'infantile dream minds'—certain episodes of the submerged ongoing interrelation with the not-known force their way now and then into our consciousness. For days and weeks afterwards, such an eidetic imprint can seem more intensely real than anything occurring to us while awake. Jung in a fine passage remarked on the lasting thrill of the special dream:

> Not all dreams are of equal importance. Even primitives distinguish between "little" and "big" dreams, or, as we might say, "insignificant" and "significant" dreams.... Significant dreams ... are often remembered for a lifetime, and not infrequently prove to be the richest jewel in the treasure-house of psychic experience. How many people have I encountered who at the first meeting could not refrain from saying: "I once had a dream!"...[6]

Even though we may not understand the experience, there is about it an electric quality of command, value, and personal import that makes it unforgettable. It was not to everyone or even to a few that this wonder occurred. The startling fact is that no one else in creation will have the identical vision experience. It is as though we were singled out and awarded a mark of distinction, a finely cut topaz or diamond quite unlike any other.

These singular moments when the mind spontaneously opens to a new vista of meaning have long been experienced and wondered about by mystics, poets, philosophers, novelists, shoemakers, and, yes, even psychologists. (Again, evidences of these outbreaks of the wilderness mind, we shall examine closely in another essay in the Clair Studies.) As we know, however, the rich, multitiered 19th-Century mind-study was co-opted early in this century by a European elite and narrowed to a 'city science.' With rhetorical *brio* and not a few sneers, Sigmund Freud would have us believe the mind's struggle with the not-known to be "nothing but" mere infantilism. Carl Jung, however, insisted that we pay attention to the rumblings of the *collective* Unconscious—as he termed the fields of intelligence we move in. And here, he was much closer to the heart of the problem. His cast of archetypes who are trying to relate with us seem a curious reprise out of antique mythologies. But archetypes, myths, or "bad dreams," these are all examples of nature's determination to catch our attention in any way possible. With astonishing inventiveness, she would bring our mind back to its main job of interfacing with the not-known. If we are ready, however, to pay attention—to allow ourselves to be fascinated, captivated by the implications of the not-known at hand—Jung's generalized figures looming out of peasant and early cityculture mythologies become somewhat anachronistic.

The vivid super-dream, or vision—that encounter with the not-known and the influx of new perceptions which followed—was referred to by the pre-Freudian psychologists as a *hypnogogic*, or sleep-teaching experience.[7] The special character of such an image play comes about when one's attention is drawn with great force and charm to some particular aspect of the complex image one is both witnessing and becoming involved in. This detail is often magnified and imbued with a remarkable perfume-like fascination. In addition to focusing us sharply on a specific, riveting symbol-metaphor, the drama may also imprint our memory indelibly with a single word or brief phrase somehow linked to the focus of the image. Later, we may see that this cryptic 'message' is a play on words of a cunning uncommon to our waking mind. From the jewel-like vividness of such a hypnogogic experience and its purposive, didactic distortions, one can imagine where some of the Eastern icons with their brilliance and their stylized exaggerations gained inspiration.

For weeks after a powerful hypnogogic experience our mind can be almost entirely possessed by the heightened reality which pervaded these nonlinear, non-temporal events. Although they were purely psychic and shared by no one else in the world, they seem so genuinely real that our waking life, centered as it is on tomorrow, becomes a fiction. The impact of the hypnogogic experience can be particularly disturbing. We might, for example, be left with the memory of having used a borrowed key to let ourself into a great house. Every moment we are seized with a fear of being discovered. Or, we must live with the uncanny, clammy memory of burying a body—no one must ever uncover it and guess the hideous secret.

Shaking off graphic feelings of hidden villainies will be very difficult. Even the waking mind is yanked back, again and again, to the eidetic vision and its terror. Gradually, one may begin to perceive that the 'super-dream' was no accident and its lasting impact no delusion. In some uncanny way details of the vision form an elaborate metaphor, the message of which is pointed and now impossible to ignore. A larger field of intelligence, somehow intimately related to us, is conveying responses to questions we had covertly, half-consciously been asking. What do the dreams *mean*? Possibly the interloper should discover his own self. And, like the earlier image in the coffin, the hidden corpse was a warning of a maturity frustrated. It is that part of the mind that, as cityman, we steadily discount. Day by day, with our focus on tomorrow and the familiar, we kill in ourselves the organic process of psychic metamorphosis. About our every scenario for the 'better tomorrow' hangs the ghastly stench of death.

On the one hand, we are creatures with a preadaptation to language and syntax. But it ought to be even clearer that we are intended to be in a continuing relationship with the unknown. That our minds should be engaged in ongoing inquiries is critical. Without that we are young trees pulled up by the roots. It is true that the field of familiarity which cityculture establishes for us is quite large. And we can accommodate ourselves within that alien sphere for a long while, perhaps our whole life. But that kind of existence is like coasting down hill on an empty tank of gas. When you have used up your initial momentum, you are dead even though you still breathe. Familiarity is not enough to maintain us as living creatures.

SLEEPING ON THE DIFFICULT QUESTION

DEEP IN THE FOLK MIND is the idea of sleeping on a problem. We must make a decision whose effects will last for years. Our mind locks into a stalemate when we try to weigh the consequences of several choices. The attempt to determine even a plausible, let alone the best, achiever scenario has us at our wits' end. We have searched the self-help books, sought out experts, but none sees the largeness of the problem as we do. All ordinary avenues of problem-solving in the cityway, through recognition and use, fail. We are on our own with no idea of how to improvise or access radically new and larger perceptions.

With our choices exhausted, we feel like a mountain climber clinging to a mere rugosity and losing friction. We go to bed, still deeply troubled, staring into the dark. At last we can no longer *think*. We give up involuntarily, our mind totally exhausted, even paralyzed. We shall have to allow fate to do with us as it will.

When we wake in the morning, however, we may feel different in some strange but pleasant way. The problem does not seem nearly so important. Or, a simple response to the difficulty has appeared whole in our minds. In our sleep something happened to us that adjusted us slightly, made us a more capable, resilient person than we were when we went to bed. This resort to sleep at a time of decision is such a commonplace that we often put off the importunate salesman with, "I think I'll sleep on that."

For an achiever, even in jest, to default a serious challenge and leave the decision to a mind asleep seems most odd. Certainly, it poses an affront to the achiever ethos which we know. Because of its incongruity, this offhand remark of "sleeping on it" suggests the survival of some tradition much older than cityculture. And its persistence seems to reveal our own deep-seated mistrust of trying to manage tomorrow with familiarity patterns that may have created the difficulty in the first place. Instead, something new, something radically new is needed. And we have this abiding sense that from out of sleep may come the help we need. The tradition of resorting to levels of the mind of which we are now little aware seems long standing.

Where this notion of sleeping on a problem might have come from was uncertain until fairly recently. It was discovered earlier in this century that Sumerian priest-kings (c.2500 B.C.) were accustomed, curiously enough, to spend a night now and then in the city's high temple. (This sanctuary atop an artificial, man-made hill was thought of as the center of the world. It replicated the archetypal sacred cave in the mountains to which shamans and prophets, from time immemorial, had resorted for experiences with the unfamiliar.) On occasions of great difficulty, such as a drought, the wisdom and help of the god or goddess (the hypnogogue) of the temple would be sought by the priest-king. After his prayers, he would lie down to sleep, anticipating a night visitation (i.e., hypnogogic experience). Significantly enough, it was only when the supplicant's mind lapsed into a non-thinking, non-achieving state that the god—or, master of the unknown—could or would convey insights and suggest new directions.

Gudea, the *ensi* of Lagash, noticed one spring that the flow of the great Tigris was not increasing as it should. Ordinarily, its level rose on schedule, year after year. An interruption in this marvelous regularity was akin to having the sun appear late one day. To the large communities of peasant cultivators in the lower delta known as Sumer, this interruption of a familiar pattern, especially in the volume of water needed for irrigation, was ominous. It could only mean that the mystery controlling the river was angry with the people of the delta. Gudea took the proper measures. He retired to the temple in hopes of instruction. And, the god of the river, Ningirsu, vouchsafed him a powerful vision. The great winged god rising from the waters commanded Gudea to rebuild and enlarge his temple.[1]

What was 'revealed' to King Gudea about the measures needed to assuage the god seems to us, of course, more contrived than a genuine, unexpected revelation. But theatrics, dissembling, and other manipulations of the populace by city leaders have always been essential to the kingship craft. What is significant for us is to learn of the survival of traditions which even in those early citycultures were most ancient, sacred, and accepted. This practice of approaching the not-known and of seeking through a special, non-sensorial faculty of the mind—that is, while asleep or during a suspension of achiever will—what the waking, active mind could not grasp may be as old as man himself.

These openings to wisdom and insights from the not-known have always been so startling they seemed 'thefts' of divine magic. Among the ancients they were thought the exclusive business of divine beings who had not forgotten man. The most well known of these Robin Hoods of the early pantheons was Prometheus who paid dearly for stealing fire from heaven. Not so well remembered are a cast of minor divinities and heroes (the so-called technic divinities, both male and female) who performed similar tasks of bringing to man insights into the intelligence that lay all about waiting to be bumped against and discovered.[2] (We have elsewhere noted that many of these special minds would have originated in the hunter world experience.) So important were these godlike episodes—when perceptions beyond yesterday and beyond tomorrow were poured forth on men—that every sort of experiment would later be made to replicate the discovery experience when desired.

Early man as hunter understood, we can believe, what is difficult for us: that perception of the uncomprehended is possible. But such perception is plainly a problem beyond the reach of the human senses. These latter faculties and their inherent limitations come into play automatically when man is awake and functioning within his culture and its familiarities. Whatever is apprehended by these ports inevitably falls within the field of the already experienced. To anything beyond the known our sensory perceptors are, by their very nature, nearly blind. The grasping of an incommensurable which has challenged man must be an exercise of the mind outside of words, outside of the everyday senses, but in real time. To experience the nature of the uncomprehended—or at least that part with which he has collided—the mind would have to grasp it all at once as a new entirety, or image. Early man knew quite well that his mind, when pitched at a certain level, or "feel," was able to move beyond the narrow boundaries of senses and words. He then might enter into whole-image games with the unfamiliar.

The so-called primitive may not have achieved gunpowder and moveable type, but the more we find out about these people, early and late, the clearer it becomes that they grasped the nature of their mind as several discrete activity levels. They also understood and experimented with the properties of each of these modes. The early hunter accepted the whole mind; it was the peasant who began to fear the dream-interface with the unfamiliar. Nevertheless, the prophet—as the shaman before him—

remained the man who could access now and then the mysteries of the uncomprehended.

From very early times every effort was made to produce on demand what originally had been an occasional, spontaneous access of fresh intelligence, or "grace." Attempts to provoke the dream-discovery mind into achieved deliveries of insights included the use of hallucinogenic herbs, mushrooms, and tobacco. These together with other kinds of intoxicants—such as music, singing, drumming, fasting, lack of sleep, prolonged chanting of mantras, ecstatic dancing, staring at a near object, blanking the mind, controlled breathing, each of these singly or in combination—might be used to blur the impingement of the external world. Similar to the shaman's retirement to a cave, these various intoxications and hard austerities denied the senses and tended to throw the mind into what was commonly called a "trance," or that quasi-sleep image-play condition we may now refer to as the theta level.

In the Old Testament we find repeated mentions of the resorts by prophets and 'sons of god' to the active dream condition, or 'ecstasy,' or "power of the Lord." Three kings marched on Moab, but ran out of water. They sought help from Elisha:

> 'As the LORD of Hosts lives, whom I serve,' said Elisha, 'I would not spare a look or a glance for you, if it were not for my regard for Jehoshaphat king of Judah. But now, fetch me a minstrel.' They fetched a minstrel, and while he was playing, the power of the LORD came upon Elisha and he said, 'This is the word of the LORD: "Pools will form all over this ravine."'[3]

Samuel anointed Saul and sent him to a community of prophets:

> When you reach the Hill of God ... you will meet a company of prophets coming down from the hill-shrine, led by lute, harp, fife, and drum, and filled with prophetic rapture. Then the spirit of the LORD will suddenly take possession of you, and you too will be rapt like a prophet and become another man. When these signs happen, do whatever the occasion demands; God will be with you.[4]

But the sophisticated of the cities sneered at these 'excesses' exhibited by the less citified. David was a strong man of the hills, a brigand captain whom the city powers of the coast, the Philistines, found easier to buy off than defeat. His first wife, Michal, Saul's daughter, seems to have been more attracted to the orderly ways of city life than to her husband's shamanistic, near-naked frenzies.

> David, wearing a linen ephod, danced without restraint before the LORD. He and all the Israelites brought up the Ark of the LORD with shouting and blowing of trumpets. But as the Ark of the LORD was entering the City of David, Saul's daughter Michal looked down through a window and saw King David leaping and capering before the LORD, and she despised him in her heart.[5]

Drugs, tricks, and sacred temples on man-made hills were mere mechanical means to duplicate what had come upon the early hunter naturally. For all the efforts of some men to approximate the trance condition and to mimic wilderness visions, we should not be diverted from that which is genuine and lies at the core of this phenomenon. Man knew long ago that his mind was more than an extension of his physical senses and more than what they recorded as familiar. He was aware early on of the mind's power and inclination to play with what was uncertain, unfamiliar, but fascinating. Once his mind fell into a special wordless attitude and became absorbed with pure alertness and curiosity, its imagemaking might reach well beyond what was thought familiar. But this attitude was not easily realized; it did not respond to desire or determination.

How to access the special mode which unlocked the radically new has always been the mystery. Where chemicals were eschewed, prayers, chants, telling of beads, limited breathing have been other resorts. (Early in the last century, one still might have come upon the primeval hunter in our Far West "crying for a vision.") It has always been hoped that when the senses of the body (which are the organs of the familiar) were asleep, then the mind might be ready to play with the not-known. Long and ancient is the tradition of men making efforts to slow their minds down from the active, past the contemplative, and into something approximating the interactive dream-discovery condition.

The self-induced trance is not without some danger and much frustration. Simulation is never the real thing. The effort to make access to the unfamiliar available on demand is one more attempt of the achiever will to maintain its continuity and dominance. Having intellectually located the source of insight and conceptualizations does not render mystery a commodity. The reduction of the iconographic, or image-play mind, to a familiar goal to be sought and achieved cannot produce an imagery of the unfamiliar. The state of mind one realizes in this manner will have been contaminated from the outset by the achiever-tomorrow notion. And whatever one experiences will most likely be the achiever mind in still another guise, this one perhaps the most seductive and difficult to detect. Here the achiever will extends its iron fingers into the living center of man's mind. Instead of restoring the individual to creation's round of wonder and surprise, achiever thinking deadens the autonomous activity of real time.

Cityculture has fairly overwhelmed all aspects of our waking mind. Our education justifies tomorrow. But we experience not insight. Avalanches of information, gorgeous entertainments, and work routines fill the 'void' between yesterday and tomorrow. We are awash with facts and charts, but the world is upside down. What was familiar has become unreliable. Mere knowledge is worthless and one must turn to strangeness. It is only man's wonder that holds steady. The drugs and ecstatic dancing used since antiquity to simulate a contact with the discovery mind are anachronisms today. All our considered efforts to affect, to imitate, to mimic, to approximate will not bring to life the wilderness mind. They only fashion another clone of cityculture.

The route to discovery lies straight into the unmarked wilderness; the mind-boggling moment of incongruity restarts the metamorphic cycle. In the acceptance and pursuit of some very difficult question that dogs us—a paradox which lies outside of cityculture's purview—we are closer to reality than in our efforts to fashion recognizable responses out of the familiar. Jung said it well: "In all chaos there is a cosmos, in all disorder a secret order."[6] The secret order will not be revealed, however, by the symbol-literate, or word-using mind. Only the image-play mind, stimulated to a white heat of wonder by a 'stoppage of time,' can in the discovery cycle's third phase grasp large chunks of what had been beyond the perceptible.[7]

BUILDING PATHWAYS AND LADDERS to the 'stars'—the concept of a step by step transformation of here into there, of resource into product—is, nevertheless, most seductive. Man has been characterized as a tool-using animal.[8] We might say instead that modern man is nothing if not a symbol-using animal. He uses metaphors as earlier man used flint scrapers and axheads. The achiever is forever raising ladders, his symbol-words being the rungs. Even Carl Jung could talk one moment about unknowns emerging from the Unconscious but at another about the marvel of "directed thinking":

> Directed thinking or, as we might also call it, *thinking in words*, is manifestly an instrument of culture, and we shall not be wrong in saying that the tremendous work of education which past centuries have devoted to directed thinking, thereby forcing it to develop from the subjective, individual sphere to the objective, social sphere, has produced a readjustment of the human mind to which we owe our modern empiricism and technics. These are absolutely new developments in the history of the world and were unknown to earlier ages.... The secret of cultural development is the *mobility and disposability of psychic energy*. Directed thinking, as we know it today, is a more or less modern acquisition which earlier ages lacked.[9]

For many of us Jung has been a mentor figure. To come upon passages containing such unresolved contradictions is a bit embarrassing. But our grand old man did what he could with what he had at hand. The history we now have of man's mind is far more complete than he knew. The individual, in the passage above, has gained "directed thinking" which equips him to build ladders, but has lost his "individual sphere." His subjective mind, which allowed him to think in unique images (often deprecated as 'fantasies'), was declared obsolete. For that loss, however, he supposedly gained the "social sphere." That a person could be both individualized and *community*-oriented at the same time—as was the ancient hunter—posed a reality which the social scientists and psychoanalysts of the early 20th Century found difficult to grasp.

The "readjustment of the human mind" Jung described as a shift to the "social sphere." What was really going on was 'socialization'—that is, adaptation to the demands of the emerging kingship reality. And, "directed thinking" meant "thinking directed." With the bulk of his mental processes (*psychic energies*) encapsulated now in words, particularly *written words*, the individual has lost control of his own reflective mind. He can only shift around, as playing with tinkertoy parts, the words and concepts supplied him by the controllers of the social sphere. He will not add anything new, for his individual sphere had been exploded. "Mobility and disposability of psychic energy" are held to be the secrets of "cultural development." At first glance such an 'advance' seems to redound to the individual's benefit. Here in this life, Jung believed it so.

Man's psychic energy, however, is his condition of being alive now, occupying the narrow path between past and future. The culture, on the other hand, is a creature of yesterdays and tomorrows. And, it can only live by its expropriation of today—that is, man's psychic energies. Every culture exists to reproduce itself, and not for the individual's benefit. Whether it be an antique peasant-priest culture or a late industrial-kingship cityculture, the business of each culture is to shape the individual to its own peculiar ends. "Mobility and disposability" seem to describe the individual's expanded use of his own potentials. What we have, rather, is 'extraction and use.' The individual has lost control of his psychic energies—which reside in the present, or "subjective, individual sphere." These are presently siphoned off through various achiever ruses by our kingship culture for its own purposes. For this extraction, or 'milking,' the resident culture returns to the individual a promissory note for a tour through fantasy time: either the paradigm yesterday or the better tomorrow. (Here we have the slick magazines, TV, Disneyworlds, and the consumer culture as America's quintessential Omega.) The individuals who are products of these "absolutely new developments in the history of the world" are—what else?—mere clones of the culture.

FINDING A PHYSICAL PATHWAY into the unfamiliar has been an obsession in every age. The hunter loaded his gear onto dog sledges and followed the bison and mastodons. Later, others got into ships and fared forth for Golden Fleeces. Our myths speak of overhearing the password to Ali Baba's cave; of inheriting Elijah's scroll-like cape; of Jack climbing the Beanstalk. We come at birth with a healthy interest in all that lies beyond symbols.

Some have sought to penetrate the unfamiliar by resorting to the familiar. They would gather all the known occurrences of a phenomenon and carefully classify them. Then, from this inventory, so went their reasoning, theories might be created which would explain the phenomenon and make it useful to man. Francis Bacon, early in the 17th Century, was one such pragmatist who excelled in expatiating on whatever was stirring his age—namely, gunpowder, books, the compass, and the very idea of invention itself. But like so many who appear capable on the surface, Sir Francis was of dwindled capacity to grasp unseen historical processes. We learn from him much about collecting details, but nothing about seizing all at once the soul of a strange situation. Bacon championed the straight-ahead attack on the not-known. And indeed, the mounting of such a campaign to secure nature's wonders seemed most sensible in those times. Consider Raleigh's efforts to study maps, interview explorers, build ships, and voyage to the Guinea coast of South America to find the riches of El Dorado.

Early in his life Bacon was struck that a few key inventions had stood the known world on its head. New cannons created new empires; moveable type democratized the exchange of ideas; the magnetic compass made the heavens, continents, and oceans seem familiar. (Bacon failed to mention that fourth wonder: the keeled ship.)

> These things and others like them lay for so many ages of the world concealed from men, nor was it by philosophy or the rational arts that they were found out at last, but by accident.... No preconceived notion could possibly have led to the discovery of them.
>
> There is therefore much ground for hoping that there are still laid up in the womb of nature many secrets of excellent use, having no affinity or parallelism with anything that is now known, but lying entirely out of the beat of the imagination, which have not yet been found out. They too no doubt will some time or other, in the course and revolution of many ages, come to light of themselves, just as the others did; only by the method of which we are now treating they can be speedily and suddenly and simultaneously presented and anticipated.[10]

"They were found out at last, but by accident.... No preconceived notion could possibly have led to the discovery of them." Here Bacon was eminently right. But then, instead of pursuing the "accident" by which these were revealed, he turns and would storm heaven with his own tower of Babel.

Bacon urged an exhaustive assembly of phenomena—plant, animal, mineral, and the products of man's hands and mind. The green and pleasant land would become a New Ark by the Cam, a latter-day Alexandria on the Thames. Out of the vastness of this material Bacon proposed a digest which would yield "Instances" and "Exclusions" from which a "Vintage" of "Inductions" might be realized. From this means the new inventions and uses of nature might be achieved with much the same regularity that one would gather timbers and build ships. Such a philosophy of forthright attack on what is strange and perhaps profitable won many converts. It seemed like a great advance in man's use of his mind. The Greeks had observed the natural world, but superficially. Their conceptualizations were biased by their philosophies and were never compared again with nature for reevaluation. For nearly 2,000 years afterwards other men made deductions from a flawed Aristotle rather than directly test hypotheses against the facts of the real world for themselves.

Bacon rightly thought it time for a fresh look. But he believed the mind of 'civilized man' quite capable of perceiving the unknown from a collection of the known. No one in those days, of course, conceived of the enormity of the world and the many difficult unknowns that such an infinitude implied. Bacon's simple induction was rather like raking up a pile of leaves to determine the meaning of a tree. Even today, in the popular mind, this Baconian experimental-inductive notion—collect evidence, draw conclusions—is thought to underlie all of our present scientific studies whether in chemistry, medicine, physics, or biotechnology. The laboratories in our great institutions are beheld with no little awe as the 'cutting edge.' We believe as an article of faith that within those confines new tomorrows for our 'American Dream' are being shaped out of strange marvels which have come solely from sifting the phenomena of nature.

The Baconian method relied on impartial, precise observations, followed by conceptualizations that would reach into the unknown. Man's *achiever* mind, however, is no such instrument; it is wholly a creature of yesterday and tomorrow. It can detect only what it happens to be familiar with already. The conclusions and concepts it can arrive at are, therefore, limited to those which will conform to what has been or what will be recognizable tomorrow.

Whatever is unknown is not a party to either the past or the future. It exists in that state of time which lies between. And it is just here that the achiever mind of cityman breaks down. Being sensitive to incongruities that deny the familiar and dealing with fields of intelligence beyond the known require faculties other than those of recognition and use which are characteristic of our conscious 'thinking' mind.

Today's successful scientist is far closer to the prophet and the shaman than he is to Bacon's model of the achiever approaching the unknown step by rational step. Nevertheless, the pragmatic pay lip-service to the inductive idea. Grants from foundations and the Government will be given to those institutions with the most elaborate plans for gathering data and making surveys. All this is safely within the overall achiever ethos. Important monies are not going to those who will have nothing to show for their efforts except what seem charming, but outrageous guesses. Hard evidence leading to no conclusions often triumphs initially over the breathtaking insight that later creates its own field of verification.

What real advances have come in our understanding have invariably been through *leaps of imagination*. In all directions the citymind is blocked: nothing works; then, a promising idea suddenly comes in out of the blue. While our science heroes can nowhere be found dancing ecstatically before the Great Unknown or calling for a minstrel to invoke the trance experience as of old, they have resorted to something similar to free their minds from time as memory and as expectation. They 'brainstorm' for suggestive ideas, even wild ones, about the nature of a paradox. They style these guesses 'calculated estimates' and 'hypotheses' and proceed to try to find the evidence that will support their guesses. If the new model arrived at by the reflective mind appears to explain the phenomenon both closely and simply, it is said to be an "elegant solution." An especially outstanding resolution is "artistic." Thus, we find the serious scientist opting for an approach to the unknown not through the nuts-and-bolts achiever method of assembly, but by semi-magical, poetical overleaps of "fancy." It is simply not in the nature of things for the unknown to yield to the familiar. (Nor will water run up hill.) Each exists in a different time. If one is to feel out the contours of the not-known, one has to resort somehow, even a little, to the *image-play* resources inherent in his non-achiever intuitive mind.

MAN MAY ALIENATE HIMSELF from today and call achieverism his personality. But nature does not readily give up on what she has started. Man is everyday teased and toyed with by what arises out of the unknown time he cannot perceive. His plans for tomorrow are disrupted. His expectations are thwarted. He forgets where he put his keys. He cannot account for his sense of defeat moments after he has succeeded. If the universe is just, why should the Indians in the Americas be driven from their lands and the Ethiopians and Sudanese starve? Anomalies, puzzles, annoyances, one after another dog him day after day. Many of these he shrugs off—what can he really do about them?

Now and then, a few of these difficult questions inevitably catch at his mind and lock it for a moment or longer into a kind of paralysis. Such an intersection of the known with the not-known results in a riddle that cannot be answered nor dismissed. Man's natural thought processes are perfectly equipped to handle such confrontations. Evolution provided a 'fall-back' to the theta-trance, or iconographic mind. Importantly enough, no minstrels or drums or drugs are needed. Any incongruity will to some degree throw the busy mind off its stride and freeze time. The Japanese *koan* ("What is the sound of one hand clapping?") is an incongruity contrived to stun the mind of the adolescent—swollen with its proud new erudition and mastery of words—and to commence his real maturation by reintroducing him to other levels of his own psyche. For an instant or longer, in this experience of bewilderment, there is no tomorrow and no yesterday. Now all hangs suspended on the iron hook of the question, the impossibility. Without familiarities these two time-impostors vanish into thin air. This forced turn to silence and the timeless allows the slower waves of the mind to rise and fill the emptiness. For it is the image-play state of the mind, and that only, which is able to consider whatever stymies our busy, achiever mind based on the senses, familiarity, and achieving the better tomorrow.

Cityculture holds that the mind works best as a linear accelerator. Man's thought is pictured as an electronic bullet moving swiftly on course to strike down his trophy. But what is the trophy of tomorrow when creation offers far greater rewards? All around us creation-nature moves in spirals. Each revolution is a metamorphosis producing endless variations of the essence that is changeless. Each rotation brings forth its own 'increase of nature.' Each cycling, or round of the seasons, guarantees a new release of *power.* In man himself this metamorphic process works when his mind responds to the unknown's attempts to engage him in its game-playing with incongruities. The vertigo that he will feel, which can range from a trace to something causing him to lie down, is indicative of the commencement of the process. The achiever-beta mind has been disabled. All ideas of yesterday and tomorrow are invalidated. Time is suddenly reset to zero; the hands of the clock do not seem to move. The image-theta mind has automatically been called in to take over and deal with this 'intrusion' of the not-known. A walk, a light nap, or some routine work may be sufficient to bring forth a 'new idea.' If not, one may have to live with the queasy feeling of uncertainty for days, weeks, even years.

The fall back into the image mind may result in an insight, or a sequence of them. A heroic response might be a grand hypnogogic tableau-vision early one morning just before waking. But it might also be a series of micro-flashes during the day, each preceded by a faint, telltale signal to pay attention, that something is about to happen. The response, however, might also be oblique and mysterious. Your hand reaches for a book on a library shelf. Something about the one next to it catches your eye. You open it and find material that fits like a jigsaw piece into the puzzle you have been trying to put together. The image mind has strange powers of directing and providing for us, once we have responded to its lures. As long as man will allow the symptoms of faintness and uncertainty which arise from contact with the not-known, he will live with riches that are golden, yet cannot be stolen. For he is then realizing his own metamorphic life process. In ways wholly unpredictable, his mind will experience one 'increase of nature' after another.

Every man represents a special entry into the not-known. Put otherwise, he represents the opportunity for the not-known to irrupt into the familiar. That the mind is continually being tantalized with questions, frustrations, and incongruities which the culture cannot 'explain' strongly suggests that each man comes with a preadaptation to some world-class puzzle.

Being overwhelmed with insight and joy is the hallmark of the wilderness mind in its latter phase. The elegance of the unexpected apprehension can be ravishing. Examples are to be found everywhere in literature, in the annals of scientific discoveries, and in the memoirs of mystics. All speak of the magical capacity of man's mind, when not clamped in the irons of the achiever mode, to experience wonders which only seconds before were incomprehensible. The 19th Century represents the finale of western man's direct experience with the planet's wilderness. Here, his exploration, conquest, and plunder of it approached a grand climax. In that astonishing century—which we shall examine in *Life Of The Mind*—some men also discovered and confronted their own personal wilderness, the unfolding wonder of which they were the living embodiment. It was a century of philosophers and poets, scientists and novelists, wilderness wanderers and delvers such as not seen before. With the passing of absolute wilderness, we have also experienced a diminution in the presence of the wilderness man, the man for all times. The subsequent cloning of the universal citymind and the projection of the better tomorrow against the clouds of heaven, it must be admitted, have been quite effective.

According to cityculture rationale, man is an emptiness to be filled. He must be outfitted with a gameboard of familiar pathways and intersections, of explicit rules and known players. Creation, even though discounted in this city rationale, may still have larger intents for man. The continuing leakage of images and tableaux from the 'dark side' suggests more to man than an inherent emptiness. The wilderness rationale argues that each is a fullness that waits to be fathomed, a marvel yet to be realized.

❑❑❑❑

REFERENCES

WORLD TIME SYMBOL: Mayan

REFERENCES

CREATION'S POWERS WERE CONCEIVED by early man to be fourfold. First, there were the four directions—east, west, north, and south—which the sun's rising and setting described. And from each of these remotenesses four great winds blew. The year itself circled from spring to spring in four seasons. Time manifested as an unseen, but undeniable, metamorphic power which ever and again appeared to function through four phases. In our Bible we find a survival of this considerable insight: Eden was itself demarcated by four rivers. The symbology of the *four* evangelists was made much of in the Middle Ages (for example, see Jung, CW, v 12, fig. 109). The Greek philosophers described creation fashioned from the four elements—fire, earth, air and water.

Early man reflected on the multiple flows of time and saw a mystery; it everywhere moved, but was essentially unchanging. The earliest notation that we have found of this conceptualization is the swastika—the very image of the vortex, chakra, prayaga, or whirling confluence. The Greek and Christian crosses seem little more than latter-day adaptations—we might even say vulgarizations—of this most ancient divination into the nature of time made by men who once lived intimately with the livingness of the earth. It was in the early kingships and citystates of Egypt and Sumer that our modern conceit of dividing and mastering time through splintering it into "useable" segments which we call hours was invented and fostered by king-priests. From then on, the reigning cultures arrogated the definition of time; they imposed canons of good and evil and yearly calendars of Holy Days and sacrifices. Individuals lost the mystery and power of time experienced for its own self.

Front cover & page ii: Spiro Mound, Oklahoma. Engraved shell gorget. Hamilton, E. K., *The Spiro Mound, The Missouri Archaeologist*, Vol. 14, October, 1952, Columbia, Mo.: University of Missouri, 1952, Plate 111.

Back cover: South India. Siva as Nataraja. In his dance of time, the great Hindu god, who depicts creation's energies, moves within a circle of flames. He holds important symbols (drum, fire, serpent) and his four arms assume significant gestures. For Western sensibilities this freakish condition of having four arms—which in Indian art was not limited to Siva alone—rivets our attention and we rarely see beyond our initial astonishment. Such an extravagance is to our minds certainly a marvel, but nevertheless 'unnatural' and superstitious. In our own frantic efforts, however, to coerce time and achieve goals, we sometimes remember Siva and wish we were so endowed. How god-like to have four hands! But what we overlook in this beautiful representation of the not-known in motion is the persistence of early man's conception of time as a dance and of its *four*-fold manifestation. The god's arms are one more representation of the original symbol of the cross or swastika. Sketch adapted from early South Indian bronzes of Siva Nataraja. (William Cohn, *Illustrations of Indian Art*, London: David Marlowe, Ltd. for the Royal India Society, 1947, Plate XXII. Also: *Ars Asiatica*, Vol. III, *Sculptures Çivaïtes*, Brussels: G. Van Oest & Co., 1921. Many views of several superb bronzes of the Nataraja.)

Page 139: Mayan. Bas-relief panel in Mayan temple at Copán. Nuttall, Zelia, *The Fundamental Principles of Old and New World Civilizations: A Comparative Research Based on a Study of the Ancient Mexican Religious, Sociological and Calendrical Systems.* Cambridge: Peabody Museum of Harvard University, 1901, pp

222-223. "The glyph which covers his face bears a native cross-symbol and this, as well as the cruciform figure, the center of which he occupies, conveys the idea of quadruplicate power." Nuttall's entire study is a fascinating effort to derive the source of the swastika form. Whether one agrees with her or not that this symbol was a representation of the precession of the Big Bear constellation around the Pole Star, her report makes a wonderful read.

PROLOGUE, p xi

<1> PDOP: Plato, *The Dialogues of Plato,* Translated by B. Jowett, Oxford: Clarendon Press, 1953, *Republic*, Sec.588.

<2> Ibid, *Republic*, Sec.580.

<3> Ibid, *Laws*, Sec.625.

<4> Ibid, *Laws*, Sec.626.

<5> Ibid, *Republic*, Sec.587.

<6> MTCB: MacLean, Paul D. *A Triune Concept of the Brain and Behavior*, Toronto: University of Toronto Press, c.1969. *passim*; Sagan, Carl, *Dragons of Eden*, N.Y.: Ballantine Books, 1977, Possible operation to correct brain 'defects,' p 213.

<7> Milton, John, *Areopagitica.*

THE ACHIEVER CYCLE, p xxix

<1> CAIW: Carroll, Lewis, *Alice in Wonderland.* N.Y.: W. W. Norton, 1971. pp 66-67.

Introductory Quotations, p xxxvi

—Hopkins, Gerald Manley, 1844-1889 *The poems of Gerard Manley Hopkins.* 4th Ed. London, Oxford U. P., 1967.

—*The Way of Lao Tzu*, translated by Wing-Tsit Chan, Indianapolis: Bobbs Merrill, 1963, number 14.

—NTSZ: Nietzsche, Friedrich Wilhelm. *Thus Spoke Zarathustra.* N.Y.: Penguin, 1961, 1969., p 46. See below, "Psychic Metamorphosis," n5.

—CATR: Coleridge, Samuel T. *The Complete Works of Samuel Taylor Coleridge.* N.Y.: Harper & Bros., 1884., p 472.

OPENING HIDDEN FRONTIERS, p 1

LIVE THE FOREST, p 4

<1> Nietzsche, NTSZ, p 78; translation cited in MWNM: George A. Morgan: *What Nietzsche Means*, N.Y.: Harper & Row, 1941, p 296.

<2> Brown, BNMB, p 385.

<3> KAOC: Koestler, Arthur, *The Act of Creation*, N.Y.: Dell, 1967, p 239.

<4> ESTE: Eliade, Mircea, *Shamanism, Archaic Techniques of Ecstasy*, Princeton: Princeton University Press, 1964. pp 38-45. Campbell, CMOG, Volume one: *Primitive Mythology*, N.Y.: Viking Press, 1959. See citation on p 252.

<5> See references below: "In Search of the Uncommon Man," notes 9 & 10.

QUARANTINE OF WILDERNESS, p 17

<1> 1 Samuel 14:41

<2> The busk, or green corn ceremony: see Corkran, David H., *The Creek Frontier, 1540-1783*, Norman: University of Oklahoma Press, 1967, pp 35-39. The Cherokees had a similar busk ceremony. Compare also with Otto's 'Mysterium Tremendum:' OITH: Otto, Rudolph, *The Idea of the Holy*, London: Oxford University Press, 1950, Chapter IV.

<3> 2 Samuel 6:5-10.

<4> For "the door we never opened:" Eliot, T. S., Four Quartets, in *Collected Poems*, N.Y.: Harcourt Brace Jovanovich, 1970, p 175.

LIVE THE MYSTERY, p 25

<1> Maugham described the power of this purposeful distortion: Philip is examining photos of El Greco's paintings. "This was not realism, and yet, and yet even in the photographs you had the impression of a troubling reality.... He felt strangely that he was on the threshold of some new discovery in life. He was tremulous with a sense of adventure.... The painter had painted exactly what he saw but he had seen with the eyes of the spirit." Earlier, Philip had been challenged by a poet: "But pray tell me what is the

meaning of life?" "I say, that's rather a difficult question. Won't you give the answer yourself?" "No, because it's worthless unless you yourself discover it. But what do you suppose you are in the world for?" Maugham, W. Somerset, *Of Human Bondage*, Garden City, N.Y.: Doubleday, 1915, 1936, pp 188, 402-404.

<2> Nikhilananda, NGSR, p 109.

<3> Otto, OITH, p 36.

THE DISCOVERY CYCLE, p 31

<1> 1 Kings, 18:24.

<2> 1 Kings, 18:27.

<3> 1 Kings, 18:36-39.

<4> 1 Samuel 9:1-14; Lame Deer, John and Richard Erdoes, *Lame Deer Seeker of Visions*, N.Y.: Simon & Schuster (Touchstone), 1972, p 151.

<5> Muir, John, *The Mountains of California*, Garden City: Doubleday/ Anchor, 1961, p 50.

<6> Ibid., pp 51-52.

<7> Jung, JCWJ, v 9.1, p 32.

<8> Densmore, DTSM, p 157; Neihardt, NBES, p 173ff; GGBF: Green, Elmer and Alyce Green, *Beyond Biofeedback*, N.Y.: Dell, 1977, p 152.

<9> Job 31:35.

<10> Job 38:2-3.

IN SEARCH OF THE UNCOMMON MAN, p 40

<1>The "technic divinities:" Lovejoy, Arthur O. and George Boas, *Primitivism and Related Ideas in Antiquity*, N.Y.: Octagon Books, 1965, 1935, p 193; Prometheus, p 200.

<2> Hoppin, James M., *Great Epochs in Art History*, Boston: Houghton, Mifflin and Co, 1903, p 14.

<3> Krober, A. L., *Handbook of the Indians of California*, N.Y.: Dover publications, 1925, 1976, pp 502-504: Yokut Jimson weed initiation of youth into dream experiences.

<4> Densmore, DTSM, pp 172-173.

<5> RPMP: Radin, Paul, *Primitive Man as Philosopher*, N.Y.: Dover Publications, 1927, 1957, p 243.

<6> Ibid, pp 245-246.

<7> Ibid, p 247.

<8> Campbell, CMOG, v 1, *Primitive Mythology*, pp 253-4. The affinity between the hunter and wilderness, or not-known, is nicely expressed by Durkheim: "So very true is it that the forest is considered his natural environment, that in a certain number of tribes, the word with which the initiation is designated signifies *that which is from the forest*." See below, Essay Five, "Closing of the Frontier," n7.

<9> Eliade, ESTE, See Chapter II, especially pp. 42f, 51f, and 60f.

<10> Ibid, p 482.

THE SOLITARY QUEST, p 50

<1> London, LCOS: Chapter XI. London came on an interesting young person, first near Oakland and again in Tahiti, whom he dubbed "Nature Man." Once given up by doctors, he had dragged himself into the woods to die. Living in a hut on a simple diet, he recovered dramatically. On returning return to the city, however, he fell deathly ill once more and was convinced to make wilderness thereafter his habitat.

<2> Nikhilananda, NGSR, "Introduction," p.32.

<3> Photograph of *Femme à la Corne* can be found in Ann Sieveking's *The Cave Artists*, London: Thames and Hudson, 1979, p 41.

<4> "Eskimo caves," Eliade, ESTE, pp 51-52. "Teton Sioux," BTSP: Brown, John Epes, *The Sacred Pipe*, Baltimore: Penguin Books, 1953, 1971, p 53. "Transcend profane," Eliade, ESTE, p 164.

<5> Campbell, CMOG, v 3, *Occidental Mythology*, London: Secker & Warburg, 1965, pp 125-140. Compare the voice out of the fire with a similar shamanistic teaching experience in the Chandogya Upanishad: Prabhavananda, Swami and Fredrerick Manchester, trans., *The Upanishads*, N.Y.: The New American Library, 1948, 1957, p 66.

<6> Exodus 3:1-6.

<7> 1 Kings 19:3-18.

<8> 1 Kings 19:11-13.

<9> 2 Kings 1-9.

<10> Charles, R. H., *Religious Development Between the Old and The New Testaments*, London: Williams and Norgate, 1914, pp 39-45.

<11> Matthew 3:4-6.

<12> Baptism not an essential Jewish rite: See series of articles on "Baptism" in *Encyclopedia of Religion and Ethics*, Edinburgh: T. & T. Clark, 1909, Vol II, p 367ff. See also: *Encyclopedia Biblica*, N.Y.: Macmillan, 1899; *New Catholic Encyclopedia*, N.Y.: McGraw-Hill, 1967.

<13> Joshua 2:9-17.

<14> 2 Kings, 1:7,8; 2:7-18.

<15> 2 Kings, 5.

<16> Sargant, SBFM, Chapters V, VI, VII; see in particular pp 207-209.

<17> Strong emotions assault the mind and produce changes: Ibid, pp 153, 155, 157, 159f, 165. IOE: Inge, William Ralph, *Outspoken Essays*, London: Longmans, Green, and Co., 1920. Inge argues that the horror of witnessing Stephen stoned, (pp 217, 218) could have provided the powerful stress that pushed Saul to the point of reversal. This classic "enatiodromia" ties directly into the etiology of conversion which Sargant has developed in his *Battle for the Mind.* See also Toynbee, TSOH, p 431: Saul the archaist becomes Paul the futurist.

<18> Kluckhohn, KLTN, pp 230-235.

<19> Bailey, Albert Edward & Charles Foster Kent, *History of the Hebrew Commonwealth*, N.Y.: Charles Scribner's Sons, 1920, 1935, p 307.

<20> Mark 1:6.

<21> Mark 9:11-13.

<22> Eliade, ESTE, *illud tempus*, pp 320-322. Eliade, EMER, *in illo tempore*, p 58; *illud tempus*, p 73. See also, *apocatastasis*, the restoration of the archetypal in the life of the individual: Jung, JCWJ, v 9.1, p 188.

<23> Eliade, ESTE, pp 320-322; Mooney, MGDR, p 777.

<24> Luke 1:36.

<25> Matthew 3:16-17.

<26> Isaiah 52:13 through all of 53.

<27> Matthew 4:1-2.

<28> Eliade, ESTE, Chapter II, pp 33-35. The “desolation” experience: LE: Laski, Marghanita, *Ecstasy*, London: The Cresset Press, 1961, Chapter XVI.

<29> Mark 1:15.

<30> Luke 4:14-22.

<31> Luke 4:28-30.

<32> Matthew, 6:26-9, 33-4.

<33> Dibelius, Martin *The Sermon on the Mount*, included in *Contemporary Thinking About Jesus* edited by Thomas Samuel Kepler, N.Y.: Cokesbury Press, 1944, p 267.

<34> Luke 17:21.

<35> Mark 12:16-17.

<36> Matthew 8:22.

<37> Many mansions, John 14:2; seek, Matthew 7:7.

<38> Amos 5:19.

PSYCHIC METAMORPHOSIS, p 70

<1> Clark, M., “A baby gets a baboon‘s heart.” [Baby Fae case], *Newsweek*, Nov 5, 1984, p 72-3. See also, *Newsweek,* Nov 26, 1984, page 94.

<2> Chakravartin, or cakravartin. See New EB, 15th Edition, “Micropedia,” vol 2, page 729.

<3> Ezekiel 1:15-21.

<4> Ezekiel 105-7.

<5> Nietzsche, NTSZ, p 46; translation cited in Morgan, MWNM, p 80.

THE CROSSOVER POINT, p 80

<1> Begley, Sharon, “Memory,” *Newsweek*, Sept 29, 1986, p 48.

THE UNFAMILIAR AT PLAY, p 87

<1> Whitehead, Alfred North, *Science and the Modern World*, N.Y.: Macmillan, 1925/1954, p 11.

<2> HECR: Hoffmann, Banesh, *Albert Einstein Creator and Rebel*, N.Y.: New American Library, 1972, p.8-9.

PHYSIOLOGY OF ACCESS, p 90

<1> DSMW: Dement, William C., *Some Must Watch While Some Must Sleep*, S.F.: W. H. Freeman, 1972, 1974, pp 13, 29. BMTM: Blakemore, Colin, *Mechanics of the Mind*, London: Cambridge University Press, 1977, pp 54-56.

<2> Dement, DSMW, pp 25-30; Foulkes, David, "Sleep," *Encyclopedia Britannica,* Chicago: Wm. Benton Pub., 1974, 1982. 15th Ed., v 16.

<3> Dement, DSMW, "REM Deprivation," pp 90-91; KSAW: Kleitman, Nathaniel, *Sleep and Wakefulness*, Chicago: University of Chicago Press, 1963, p 105.

<4> Ibid., p 106.

<5> Blakemore, BMTM, pp 50-51; Brown, BNMB, pp 364-365.

<6> Brown, BNMB, p 384. Proper mention should also be made of earlier pioneers in this field of mind study. Among them were the men affiliated with Dr. Lawrence S. Kubie and others with Dr. Trigant Burrow. The Kubie group devised a form of biofeedback in which the patient was allowed to listen to an amplification of his own breathing. In about ten minutes his mind had "relaxed" enough so that he could recall memories blocked by his more active mind. Kubie called this condition of mind—which was probably in the theta or lower range—the "hypnogogic reverie" state. (See: Lawrence S. Kubie, "The Use Of Induced Hypnagogic Reveries In The Recovery Of Repressed Amnesic Data," *Bulletin of the Menninger Clinic*, v 7, 1943, Topeka, Kansas.) The Burrow group had experimented with a form of what might be summarized as a kind of group meditation. Eye movement, when held to a minimum by practice, slowed the mind down and produced a specialized attention which Burrow termed *cotention*. From a busy, wishful concern with words and concepts—which are mere substitutes for the concrete—the mind "achieves a direct relation to the environment." This Burrow felt was man's primary, or natural mind which had been covered over by a forced adaption to the prevailing culture and use of its specialized symbols. (See Trigant Burrow, "Kymograph Records of Neuromuscular Patterns," *Psychosomatic Medicine*, v III, number 2, April, 1941.); Ibid, p 369.

<7> Ibid, p 364.

<9> Ibid, p 379.

<9> Dement, DSMW, p 32. Blakemore, BMTM, p 56.

<10> Brown, BNMB, pp 364, 379, 383. Green, GGBF, See the important Chapter 7 on "theta training" and especially p 123.

<11> Sokolov, E. N., "The Orienting Reflex, Its Structure and Mechanisms", in OREB: *Orienting Reflex and Exploratory Behavior*, edited by L. T. Voronin, A. N. Leontiev, A. R. Luria, E. N. Sololov, *and* O. S. Vinogradova, Washington, D. C.: American Institute of Biological Sciences, 1965, page 143.

<12> Brown, BNMB, p 353.

<13> BOMC: Bowart, Walter, *Operation Mind Control*, N.Y.: Dell Publishing Co., 1978, Chapter Ten; Sargant, SBFM, conversion possible when senses powerfully affected, see note 17, in "The Solitary Quest" above.

<14> McConnell, James V., "Criminals Can Be Brainwashed Now," *Psychology Today*, April 1970, p 14.

<15> Brown, BNMB, pp 2-3; Jung, JCWJ, v 5., *Symbols of Transformation*, Princeton: Princeton University Press, 1967: "dream-thinking," a regression, par. 25; "archaic thinking," "a peculiarity of children and primitives," par. 32, par. 37.

<16> Bronstein, A. I., N. A. Itina, A. G. Kamenetskaia, and V. A. Sytova, "The Orienting Reactions In Newborn Children", in OREB, p 312.

THRESHOLD OF THE UNKNOWN, p 109

<1> Sargant, SBFM, pp 119-128, but pp 1223-124 in particular. Also, Burke, Kenneth, *Language As Symbolic Action*, Berkeley: University of California Press, 1968/1966, pp 45-46. Jung, JCWJ, v 18, par's 460, 483-494. Mable Dodge Luhan tells an amusing story in her autobiography about how she made up dreams to impress and 'satisfy' Dr. Brill: *Intimate Memories*, v 3, *Makers and Shakers*, N.Y.: Harcourt, Brace and Co., 1936, pp 505-512. She was one of the first to write articles for the popular press delineating the Freudian doctrines for the common people unable to afford the high-priced 'specialists.'

<2> SDTM: Stewart, Kilton, "Dream Theory in Malaya," in *Altered States of Consciousness*, Editor: Charles T. Tart, N.Y.: John Wiley & Sons, Inc., 1969, p 161.

<3> Ibid, p 166.

<4> Stevenson, Robert Louis, "A Chapter on Dreams," *Across The Plains*, N.Y.: Charles Scribner's Sons, 1923. See also Elmer Green's experiments with "sleeping on a problem": Green, GGBB, pp 8-9.

<5> Jung, JCWJ, v 9-1, paragraphs 505, 516, 521.

<6> Jung, JCWJ, v 8, paragraph 554.

<7> Myers, Frederick W. H., *Human Personality and its Survival of Bodily Death*, 2 vols, N. Y.: Arno Press, 1975, vol 1, pp 124-125. Green, GBBF, p 120 & *passim*.

SLEEPING ON THE DIFFICULT QUESTION, p 121

<1> Jacobsen, Thorkild, "The Function of the State," in Frankfort, Henri, H. A. Frankfort, John A. Wilson, Thorkild Jacobsen, & William A. Irwin, *The Intellectual Adventure of Ancient Man*, Chicago: University of Chicago Press, 1946. Chapter VI. pages 189-191; footnote on page 201, "Entemena, Cone A," reference to Gudea's vision of the god etc.

<2> See above, the first note under "In Search of the Uncommon Man."

<3> 2 Kings. 3:14-16.

<4> 1 Samuel 10:5-8.

<5> 2 Samuel 6:14-16.

<6> Jung, JCWJ, v 9.1, p 32, par.66.

<7> Joseph Campbell calls this "stoppage of time" an "aesthetic arrest": Campbell, CMOG, v 4, *Creative Mythology*, N.Y.: Viking/ Compass, 1968, 1970, p 67.

<8> Carlye, Thomas, *Sartor Resartus*, Edited by Charles Fredrerick Harrold, N.Y.: The Odyssey Press, Book I, Chapter V. p 41. See also, Emerson, ESFE, *The Poet,* pp 227-231; the poet as energizer of symbols. Carlyle was protean. Some of his passages on time are memorable: "That the Thought-forms, Space and Time … should condition and determine our whole Practical reasonings, conceptions, and imagings or imaginings, seems altogether fit, just, and unavoidable. But that they should … blind us to the wonder everywhere lying close on us, seems nowise so…. Consider, then, with thyself how their thin disguises hide from us the brightest God-effulgences! … Your grand anti-magician, and

universal wonder-hider, is this same lying Time." Carlyle, *Sartor Resartus*, Book III, Chapter VIII, pp 262-263.

<9> Jung, JCWJ, p 16, par.17.

<10> Bacon, Francis *Novum Organum*, N.Y.: Random House, 1939. page 73, Aphorism cix.

❑❑❑❑

BIBLIOGRAPHY

— B —

Bacon, Francis. *Novum Organum.* N.Y.: Random House, 1939.

Bailey, Albert Edward & Charles Foster Kent. *History of the Hebrew Commonwealth.* N.Y.: Charles Scribner's Sons, 1920, 1935.

Begley, Sharon. "Memory," *Newsweek*, September 29, 1986.

Blakemore, Colin. *Mechanics of the Mind.* London: Cambridge University Press, 1977.

Bowart, Walter. *Operation Mind Control.* N.Y.: Dell Publishing Co, 1978.

Bronstein, A. I., N. A. Itina, A. G. Kamenetskaia, and V. A. Sytova, "The Orienting Reactions In Newborn Children." In *Orienting Reflex and Exploratory Behavior.*

Brown, Barbara B. *New Mind, New Body.* N.Y.: Bantam Books, 1975.

Brown, John Epes. *The Sacred Pipe.* Baltimore: Penguin Books, 1953, 1971.

Burke, Kenneth. *Language as Symbolic Action.* Berkeley: University of California Press, 1968/1966.

— C —

Campbell, Joseph. *The Hero with a Thousand Faces.* Princeton: Princeton University Press, 1949, 1960.

Campbell, Joseph. *The Masks of God.* 4 volumes. N.Y.: Viking Press, 1959, 1968.

Carlyle, Thomas. *Sartor Resartus*. Edited by Charles Frederick Harrold. N.Y.: The Odyssey Press, 1937.

Carroll, Lewis. *Alice in Wonderland.* N.Y.: W. W. Norton, 1971.

Charles, R. H. *Religious Development Between the Old and The New Testaments.* London: Williams and Norgate, 1914.

Coleridge, Samuel T. *The Complete Works of Samuel Taylor Coleridge*. N.Y.: Harper & Bros., 1884.

Corkran, David H. *The Creek Frontier, 1540-1783*. Norman: University of Oklahoma Press, 1967.

—D—

Dement, William C. *Some Must Watch While Some Must Sleep.* San Francisco: W. H. Freeman, 1972, 1974.

Dibelius, Martin. *The Sermon on the Mount*, included in *Contemporary thinking about Jesus.* Edited by Thomas Samuel Kepler. N.Y.: Cokesbury Press, 1944.

—E—

Eliade, Mircea. *The Myth of the Eternal Return*. Princeton: Princeton University Press, 1954.

Eliade, Mircea. *Shamanism, Archaic Techniques of Ecstasy.* Princeton: Princeton University Press, 1964.

Eliot, T. S. *Collected Poems* 1909-1962. N.Y.: Harcourt Brace Jovanovich, 1970.

Encyclopedia Biblica. N.Y.: Macmillan, 1899.

Encyclopedia of Religion and Ethics, Edinburgh: T. & T. Clark, 1909.

—F—

Freud, Sigmund. *Complete Psychological Works of Sigmund Freud.* London: The Hogarth Press, 1961.

—G—

Green, Elmer and Alyce Green. *Beyond Biofeedback.* N.Y.: Dell, 1977.

—H—

Hoffman, Banesh. *Albert Einstein Creator & Rebel.* N.Y.: Viking/ New American Library, 1972, 1973.

Hopkins, Gerald Manley. *The Poems of Gerald Manley Hopkins.* 4th Ed. London: Oxford U. P., 1967.

Hoppin, James M. *Great Epochs in Art History.* Boston: Houghton Mifflin, 1903.

—I—

Inge, William Ralph. *Outspoken Essays.* London: Longmans, Green, and Co., 1920

—J—

Jacoben, Thorkild. "The Function of the State." In *The Intellectual Adventure of Ancient Man.* By Henri Frankfort et al. Chicago: University of Chicago, 1946.

Jung, C. G. *The Collected Works of C. G. Jung*, 20 Volumes. Princeton: Princeton University Press.

—K—

Kleitman, Nathaniel. *Sleep and Wakefulness.* Chicago: University of Chicago Press, 1963.

Kluckhohn, Clyde and Dorothea Leighton. *The Navajo.* Garden City, N.Y.: Doubleday/Anchor, 1962.

Koestler, Arthur. *The Act of Creation.* N.Y.: Dell, 1967.

Krober, A. L. *Handbook of the Indians of California.* N. Y.: Dover Publications, 1925, 1976.

Kubie, Lawrence S. "The Use of Induced Hypnagogic Reveries in The Recovery of Repressed Amnesic Data." *Bulletin of the Menninger Clinic*, v 7, 1943. Topeka, Kansas.

—L—

Lame Deer, John and Richard Erdoes. *Lame Deer Seeker of Visions.* N.Y.: Simon & Schuster (Touchstone), 1972.

Lao Tzu. *The Way of Lao Tzu.* Translated by Wing-Tsit Chan. Indianapolis: Bobbs Merrill, 1963, number 14.

Laski, Marghanita. *Ecstasy*. London: The Cresset Press, 1961.

London, Jack. *Cruise of the Snark.* N.Y.: Macmillan, 1911.

Lovejoy, Arthur O. And George Boas. *Primitivism and Related Ideas in Antiquity*. N.Y.: Octagon Books, 1965, 1935.

Luhan, Mabel Dodge. *Intimate Memories*. N.Y.: Harcourt, Brace & Company, 1936.

—M—

McConnell, James V. "Criminals Can Be Brainwashed Now." *Psychology Today*. April, 1970.

Mooney, James. *The Ghost-Dance Religion and the Sioux Outbreak of 1890*. U.S. Bureau of American Ethnology, 14th Annual Report, 1892-1893. Washington, D.C., 1896.

Muir, John. *The Mountains of California.* Garden City, N.Y.: Doubleday/Anchor, 1961.

Myers, Frederick William Henry. *Human Personality and its Survival of Bodily Death*. N.Y.: Arno Press, 1975.

—N—

Neihardt, John G. *Black Elk Speaks*. N.Y.: Simon and Schuster, 1972.

New Catholic Encyclopedia N.Y.: McGraw-Hill, 1967.

Nietzsche, Friedrich Wilhelm. Thus *Spoke Zarathustra*. N.Y.: Penguin, 1961, 1969.

Nikhilananda, Swami, Trans. *The Gospel of Sri Ramakrishna.* N.Y.: Ramakrishna-Vivekananda Center, 1942.

—O—

Otto, Rudolph. *The Idea of the Holy*. London: Oxford University Press, 1923, 1950.

—R—

Radin, Paul. *Primitive Man as Philosopher*. N.Y.: Dover Publications, 1927, 1957.

—S—

Sargant, William. *Battle for the Mind.* N.Y.: Harper & Row, 1957, 1959.

Sieveking, Ann. *The Cave Artists.* London: Thames and Hudson, 1979.

Sokolov, E. N. *The Orienting Reflex, Its Structure and Mechanisms.* In *Orienting Reflex and Exploratory Behavior.* Edited by L. T. Voronin, A. N. Leontiev, A. R. Luria, E. N. Sokolov, and O. S. Vinogradova. Washington, D. C.: American Institute of Biological Sciences, 1965.

Stevenson, Robert Louis. *Across the Plains.* N.Y.: Charles Scribner's Sons, 1923.

Stewart, Kilton. "Dream Theory in Malaya." In Tart, ASC.(See next entry.)

—T—

Tart, Charles T., editor. *Altered States of Consciousness.* N.Y.: John Wiley & Sons, Inc., 1969.

Toynbee, Arnold J. *A Study of History.* Abridgement of Vol I-VI by D. C. Somervell. N.Y.: Oxford University Press, 1947.

—U—

Voronin, T. & others, editors. *Orienting Reflex and Exploratory Behavior.* Editors: L. T. Voronin, A. N. Leontiev, A. R. Luria, E. N. Sokolov, and O. S. Vinogradova. Washington, D. C.: American Institute of Biological Sciences, 1965.

—W—

Whitehead, Alfred North. *Science and the Modern World.* N.Y.: Macmillan, 1925/1954.

❑❑❑❑

INDEX

C

D

E

F

G

H

I

J

K

L

M

N

O

P

Q

R

S

T

U

V

W

Y

Z